THE ILLUSTRATED HISTORY OF MAN IN SPACE

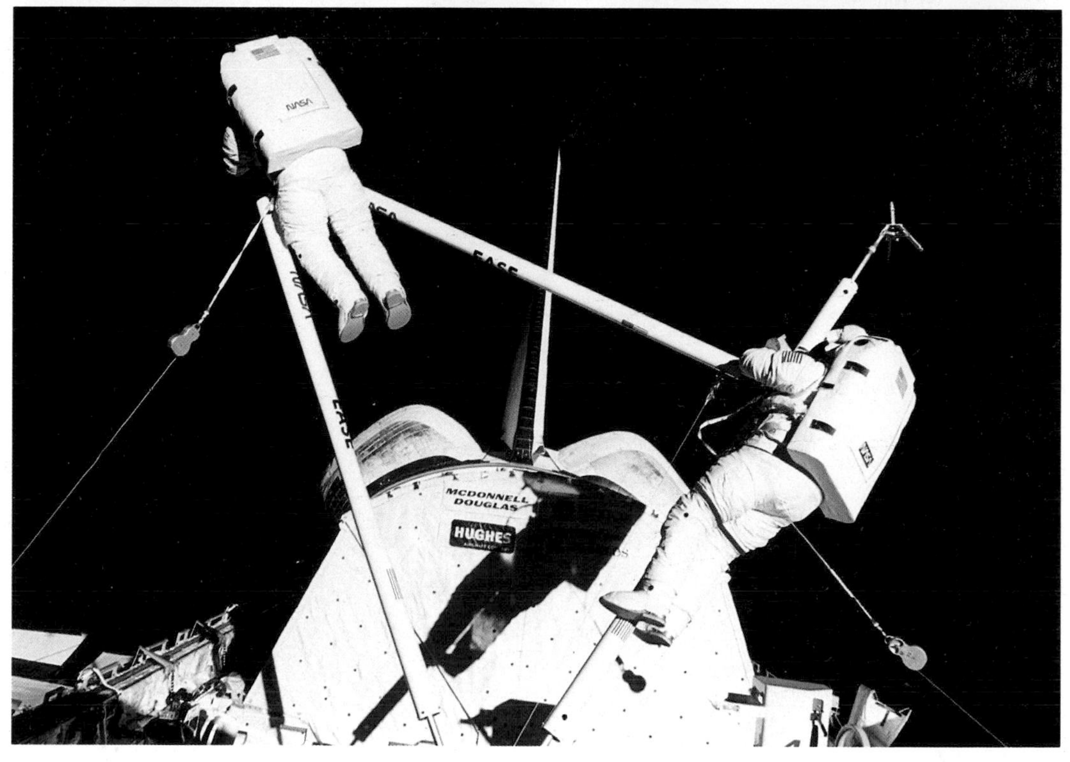

UNITED
STATES

THE ILLUSTRATED HISTORY OF MAN IN SPACE

ROBIN KERROD

This book was devised and produced by
Multimedia Books Limited

Editor: Linda Osband
Design: John Strange
Production: Zivia Desai

First published in the United States of America in 1989 by Mallard Press.

ISBN 0-792-45040-X

Typeset by BWS Typesetters, London
Origination by Imago Publishing Limited
Printed in Italy by Imago Publishing Limited

Half title: On shuttle mission 61-8 astronauts practice space-station construction techniques.

Title: Man on the Moon, with all his paraphernalia – spacesuit, spacecraft and lunar runabout.

This page:
(Top) Three of the record-breaking crew of eight of *Atlantis* on the Spacelab D1 mission.
(Bottom) Jerry Ross and Sherwood Spring show how well man can perform in zero-g conditions.

CONTENTS

INTRODUCTION

MAN MUST EXPLORE

'Po-ye-kha-li!'
The jubilant exclamation ('Off we go!') came from the man strapped inside the capsule at the top of the rocket that was beginning its fiery ascent from the launch pad. The place: the Baikonur Cosmodrome on the Russian steppe. The date: 12 April 1961. The man: Yuri Alekseyevich Gagarin.

Gagarin was setting off on a journey that no human being had attempted before; a journey into the unknown, into the most hostile environment in the universe for man – space. He went, he saw, he conquered. And he survived.

He survived the punishing g-forces created by the vicious acceleration of the launch rockets, when his body became up to eight times heavier. In orbit he survived the strange state of weightlessness, when his body organs floated free. He survived the shock and searing heat of re-entry into the atmosphere at 10 times the speed of a rifle bullet. He survived!

More than 110 times since then astronauts and cosmonauts, men and women, have followed the trail that Gagarin blazed and ridden a pillar of fire into the heavens. Today they fly into space in sophisticated space planes and undertake tasks that were once the stuff of science fiction. They go spacewalking to carry out repairs on their spacecraft; they flit hither and thither in Buck Rogers style jet-propelled backpacks, recovering a satellite here, mending one there. In permanent space stations they make their home in space for months, even a year at a time. Ever the adaptable creature on Earth, man has proved he can adapt to life off Earth as well.

Space remains, however, an unforgivingly hostile frontier that man enters at his peril. Less than meticulous preparation for running the space gauntlet invites disaster. Nothing underlined this more than the *Challenger* tragedy of 1986, when 'seven star voyagers' met their particularly horrifying and very public death in the Florida skies 73 seconds after a launch that should never have been allowed to take place.

More than anyone, the astronauts and cosmonauts are only too aware of the hazards that space flight presents. Wrote Gagarin after the death of Vladimir Komarov, the first man to perish on a space mission: 'The road to the stars is steep and dangerous. But we're not afraid. Nothing will stop us.'

But why go into space? Why invite disaster on that most deadly of frontiers? In July 1971 Apollo 15 moonwalker David Scott pinpointed the reason for it all when he stepped out on to one of the most dramatic landscapes the Moon has to offer: 'As I stand here in the wonders of Hadley,' he said, 'I sort of realize there's a fundamental truth to our nature. Man must explore. And this is exploration at its greatest.'

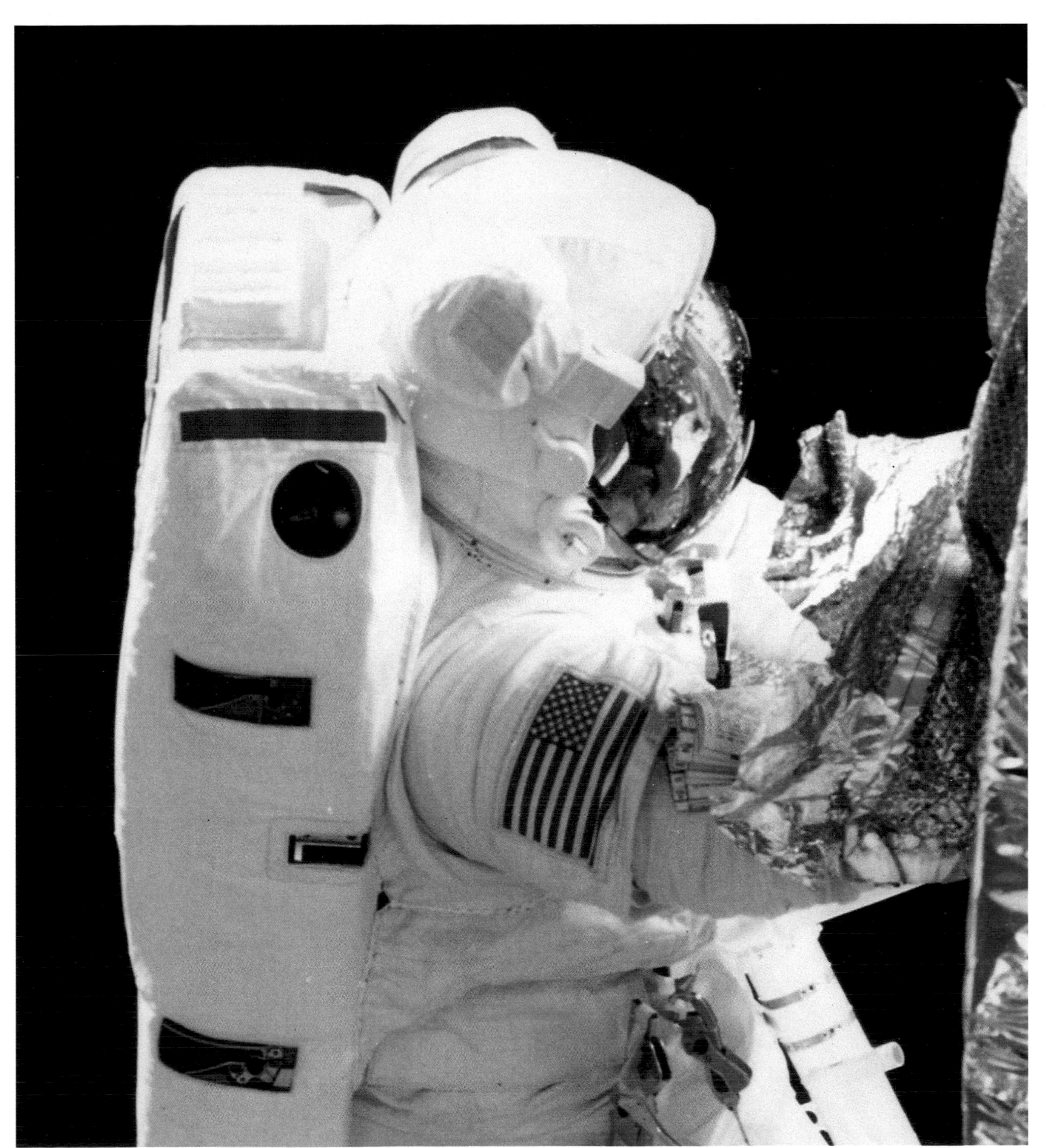

◀ **Gently does it**
On shuttle mission 51-A Joseph Allen manhandles the Westar satellite into the payload bay of orbiter *Discovery* to complete the most daring in-orbit EVA operation ever attempted in the history of space flight. Westar is the second of two comsats recovered during the mission.

Chapter 1

EXPLORING AND EXPLOITING SPACE

◀ **V2 lift-off: White Sands**
Conceived in Germany in the early 1940s, the V2 is the direct ancestor of the modern space rocket. Sixty-four V2s were launched at White Sands Proving Ground, New Mexico, between 1946 and 1952.

▶ **Earth: Oasis of life**
Weather satellite Meteosat captures a magnificent image of our home planet, sparkling like a jewel against the inky blackness of space. Satellites have revolutionized weather forecasting.

'Beep...Beep...Beep...Beep...'
On 4 October 1957 a beeping sound transmitted from a little sphere of aluminum circling hundreds of kilometers above the Earth announced to the world the beginning of the Space Age. It was the first artificial satellite, Sputnik 1. The science and technology that spawned this man-made moon can be traced directly to the notorious V2 rocket bomb of the 1940s, and ultimately to the invention of the rocket by the Chinese maybe as long as a millennium ago. Man's dreams of journeying in the heavens date back even further, to the days of classical Greece and Rome, where they are interwoven in fantastic myth and legend.

From the turn of this century, novelists such as Verne and Wells; theoreticians such as Tsiolkovsky; and practical men such as Goddard, von Braun and Korolev; all played their part in advancing the awareness, concepts, and nuts and bolts of space flight. In this chapter we trace how their dedication to a dream resulted in the flight of Sputnik and the thousands of other satellites that have followed. We examine the formidable hardware needed to loft such bodies into space in the teeth of the all-pervasive force of gravity, and look to the benefits they have wrought. This sets the scene for the main thrust of the book, the march into space by man himself.

Fully two decades before the Wright brothers even took to the air at Kitty Hawk, Tsiolkovsky set down his vision of the future. 'Man will not remain on Earth for ever,' he wrote. 'In his pursuit of light and space he will at first timidly probe beyond the atmosphere, then conquer all of circumsolar space.'

Moon shot: 17th century
Space fantasies feature in the writings of celebrated French novelist and swordsman Cyrano de Bergerac. This illustration is taken from *The Voyage to the Moon*, written in 1649.

EARTHLINGS BEGAN EXPLORING SPACE in their imagination centuries before they had any real idea of what it was like. Each early civilization had its own concept of the universe. To the Egyptians, for example, the universe consisted of a flat Earth, spanned by the heavens, which were formed by the star-spangled body of the sky goddess Nut. The Sun, the god Ra, traveled across the heavens each day in a boat. The Greek mathematician Pythagoras (500s BC) was one of the first to believe that the Earth is a sphere. Two centuries later the philosopher Aristotle cited evidence to support the idea of a round Earth – for example, that the Earth's shadow on the Moon during an eclipse is curved.

By the time of Ptolemy of Alexandria (*c*. AD 150), Greek ideas of the universe had been formalized into what is called the Ptolemaic system. A spherical Earth was fixed at the center of the universe, around which circled all the heavenly bodies – the Sun, Moon, stars and planets. The stars were fixed to the inside of a great enveloping celestial sphere. Attempts were made to explain the complex motion of the planets through the heavens by the use of secondary circles. A planet moved in a small circle (epicycle), whose center (deferent) moved in a great circle around the Earth. Ptolemy's deferent and epicycle system did not work very well, but nevertheless it was accepted virtually without question for 14 centuries.

No one, of course, had any idea of what the heavenly bodies were like. Some philosophers had suggested that the Sun, Moon and stars were made of fire. Anaxagoras of Clazomenae (*400s BC*) had even gone so far as to suggest that the Moon was made of the same stuff as the Earth. Such beliefs were heresy. (How delighted would Anaxagoras have been had he been able in July 1969 to see Neil Armstrong set foot on the lunar soil and prove that he was right!)

Whether he knew of Anaxagoras's ideas or no, the Greek writer Lucian of Samasota, in the second century AD, wrote one of the first stories about traveling to the Moon. He described how a shipload of sailors passing through the Pillars of Hercules (Straits of Gibraltar) were whisked by a waterspout to the Moon, where they met the Moon king. He called his book *True Histories,* though he admitted it was a pack of lies from beginning to end!

Traveling to the Moon

By the turn of the 17th century Ptolemy's idea of Earth-centered universe was being challenged by the Sun-centered, or solar system that Copernicus had put forward in 1543, though the

Church was quick to brand Copernicans as heretics. In 1609 the Italian astronomer Galileo trained his telescope on the heavens and saw with his own eyes evidence to support Copernicus's ideas. He also saw that the Moon was another world like our own.

Perhaps his discoveries inspired the tales about traveling to the Moon written by Bishop Francis Goodwin in England and the celebrated duellist and writer Cyrano de Bergerac in France a few decades later. Bishop Goodwin's hero, in *Man in the Moon,* found himself towed to that world in a chariot drawn by swans. De Bergerac suggested that to journey into space, a traveler should in the early morning attach to himself bottles of dew. When the Sun came up, the dew would vanish into the air, taking the traveler with it!

▼ **Fire arrows**
Seventeenth-century examples of Chinese rocket-propelled fire arrows, little changed for 400 years. This illustration appears in a military treatise entitled *Wu-Pei-Chih* (1621).

▶ **Lunar voyager: 19th century**
'Aerial Diligence', a form of lunar transportation powered by flying Moon men. It is a product of the imagination of distinguished English astronomer Sir John Herschel, who was convinced the Moon is inhabited.

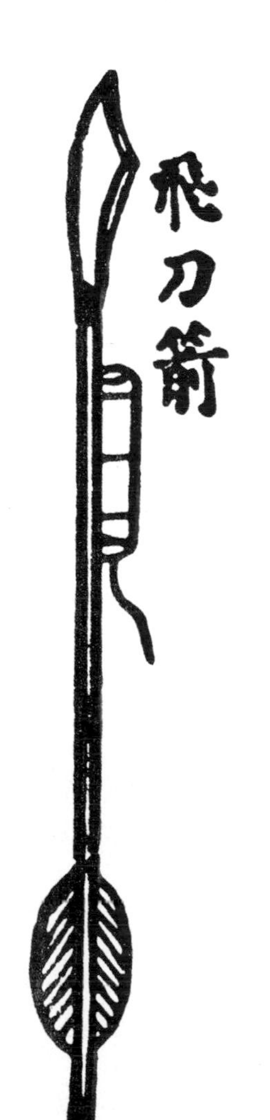

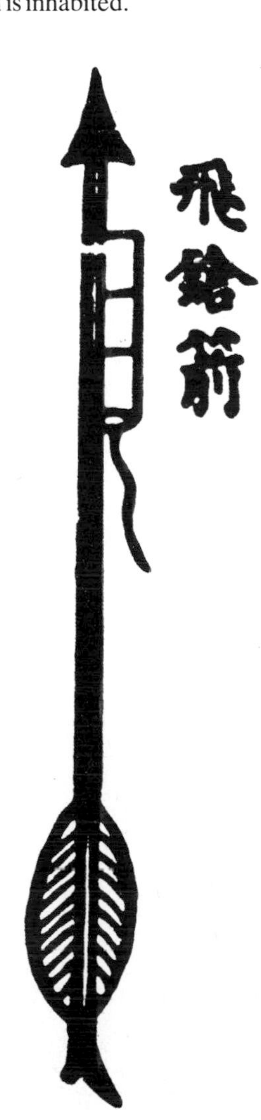

With such tales writers pioneered the art form of science fiction. In the 19th century came two skillful practitioners of the art, whose writings inspired many not only to dream of space flight but to do something practical about achieving it. First was the French writer Jules Verne, who in 1865 published quite a realistic account of a voyage to the Moon in the book *From the Earth to the Moon*. In this book and the sequel *Around the Moon,* Verne incorporated many well-founded ideas about what space flight would be like. He foresaw, for example, the problem of g-forces caused by fierce acceleration on lift-off. He calculated that a 'projectile' would have to be fired from the Earth at a speed of 40,000 km/h (25,000 mph) in order to escape gravity and reach the Moon. Prophetically he chose a launch site for his Moon shot in Florida, close to where man eventually did shoot for the Moon a century later! Verne was wrong, however, in the method he chose to propel his heroes to the Moon. They were fired from a cannon in a hollow projectile.

The English writer H. G. Wells chose an ingenious method of launching a spaceship to the Moon in his book *The First Men in the Moon* (1901). It was propelled by means of an anti-gravity material called cavorite. Exposing the cavorite towards Earth neutralized gravity in that direction, allowing the Moon's gravity to pull the spaceship towards it. But like water-spouts, evaporating dew, flying geese and cannons, cavorite is not a practical method of space propulsion!

▸ **Konstantin Tsiolkovsky**
In the late 1800s Tsiolkovsky, the father of astronautics, prophetically sets down many of the principles on which space flight depends. He foresees that man will one day reach out 'to conquer all of circumsolar space'.

▸ (opposite) **Goddard and rocket, 1926**
Standing in a snow-covered landscape on 16 March 1926, Goddard is pictured prior to the successful launch of the first liquid-propellant rocket.

▾ **Robert H. Goddard: 1924**
While studying for his doctorate at Clark University, Goddard theorizes about traveling to the Moon. This earns him the nickname 'Moon Man' and much ridicule.

Arrows of fire

Verne and Wells had both overlooked a method of propulsion that already existed which promised a practical route into space. It was the rocket. The rocket's history dates back at least to the 13th century, when the Chinese are known to have used 'fire arrows' in their war against the Mongols. They were missiles consisting of an arrow attached to a tube of gunpowder. They were probably inefficient as weapons, but were undoubtedly invaluable in spreading terror and confusion among the enemy as they whooshed and crackled overhead.

Over the centuries rockets were used fitfully in warfare but did not become an effective weapon until the early 1800s, when an English colonel, William Congreve, systematically improved them. English rocket squadrons saw successful action at the Battle of Waterloo in 1815 and elsewhere in Europe. By the mid-1800s most European armies had formed their own rocket squadrons. So had Russia, and it is in Russia that the story of space rocketry really begins.

On 17 September 1857 Konstantin Eduardovich Tsiolkovsky came into the world in the town of Izhevskoye, south-west of Moscow. By the age of 15 and deaf as a result of an earlier illness, he had developed a consuming interest in rockets, which led him six years later to suggest that they could be used to travel in space. He became a school-teacher and in one vacation in 1883 wrote a book called *Free Space* in which, among other things, he predicted the state of weightlessness: 'Any place in free space', he wrote, 'could serve as an excellent bed or an excellent chair.'

Tsiolkovsky continued to devote all his spare time to considering details of space flight and the design of rockets. As a sideline he wrote science-fiction stories which gave vent to his imagination, such as 'On the Moon' (1887). Seven years later he moved to Kaluga, 200 km (125 miles) from Moscow, where he remained for the rest of his life. It was there that he refined his theories of traveling in space which have led to him being considered the father of astronautics, the science of space flight. He published his ideas in 1903 in a paper entitled 'Exploring Space with Reactive Devices'. ('Reactive devices' means 'rockets'.)

In the paper Tsiolkovsky stated that only rockets could be used for propulsion in space, and that a much more powerful propellant than gunpowder would be needed. He suggested that a liquid-propellant rocket burning liquid hydrogen and liquid oxygen would be best. (It is, and these substances are now used, for example, in the Soviet Energia launch vehicle and the US space shuttle.) Furthermore, Tsiolkovsky realized that a space launch vehicle would have to be made up of several rocket stages, which would fire and separate in turn. He thus established the principle of the step rocket, which indeed provides the practical means of getting into space.

Reaching extreme altitudes

Quite unaware of Tsiolkovsky's work, a young man half a world away was developing his own ideas about rockets and space. He was Robert Hutchings Goddard, born at Worcester, Massachusetts, in 1882. From an early age, he became a space fanatic after reading the stories of Verne and Wells. By 1914 he was teaching at Clark University in Worcester, where he had obtained his doctorate three years before. And he already had several

▼ **Preparing a V2**
With the war all but lost, Germany launches a last-ditch offensive in 1945 with V2 rocket bombs. The aim is to subdue England; the target is London. Here final preparations are being made for a launch.

▼ (right) **Goddard's rocket range**
In the 1930s Goddard conducts many rocket flights from this range at Roswell, New Mexico.

patents relating to rocket motors to his credit. When the United States entered the First World War in 1917, Goddard went to California to work on military rockets, returning to Clark University afterwards.

In 1919 the Smithsonian Institution published a paper Goddard had written entitled 'A Method of Reaching Extreme Altitudes'. The paper set down much of his advanced thinking on rocketry and included a suggestion that a rocket be dispatched to the Moon on a demonstration flight. This idea attracted ridicule from the press and the public at large, and earned him the name of 'Moon Man'. Smarting from such treatment, Goddard fought shy of publicity thereafter. But he continued his work, and by 1926 he was ready to test a novel kind of rocket, which used liquid propellants. He had reached the same conclusion as Tsiolkovsky that solid-propellant rockets were not powerful enough for high-altitude flights.

On 16 March 1926 he lit his rocket, which burned gasoline and liquid oxygen propellants, and watched it reach a height of over 40 feet (12 meters) before plummeting to the ground after a flight of just 2½ seconds. It was a small start, but it introduced the type of rocket that would eventually lead man into space. Goddard continued to develop his rockets through the 1930s. By the time Tsiolkovsky died in 1935, Goddard's rockets, fired from Roswell, New Mexico, could travel at speeds of over 1100 km/h (700 mph) and reach altitudes of 2.5 km (1.5 miles). In 1941 he was flight-testing rockets using turbopumps to feed propellants into the combustion chamber and so augment thrust. However, in historical perspective, his efforts were being eclipsed, unknown to most of the world, by a German team located on the remote island of Peenemünde on the Baltic.

The infamous V2

At Peenemünde a research team headed by one Wernher von Braun was developing rockets for the military. They were liquid propellant rockets using alcohol and liquid oxygen as propellants. From 1937 they had been testing rockets known as A3 and A5, which by 1941 were reaching altitudes of up to 13 km (8 miles). These firings paved the way for the development of the A4 – 14 meters (46 feet) long, weighing 13 tonnes, with a thrust of 25 tonnes. There had never been a rocket like it before.

In October 1942 the first perfect flight of the A4 took place, following two failures in March and August. By 1943, after Peenemünde had been bombed, von Braun, his team and his rockets had transferred to an underground factory in the Harz Mountains called Mittelwerk. The A4 had now been fitted with a warhead and had become the sinister V2, Hitler's second revenge weapon. In September 1944 V2s were launched against England, mostly targeted on the nation's capital, London. Over the next few months some 1200 V2s fell on England and killed upwards of 2500 people. Against the V2, the world's first ballistic missile, there was no defense. It plunged suddenly from out of the heavens at a speed of 5000 km/h (3000 mph), exploding its 975-kg (2150-pound) warhead on impact.

Fortunately, the V2 assault came too late to alter the course of the war. In the spring of 1945 the Allies were advancing on all fronts. Von Braun and about 100 other members of his team gave themselves up to the advancing US Army. A year later they were in New Mexico at the White Sands Proving Ground, where they began assembling V2s from parts shipped over from Germany and carrying out test firings. In February 1949 they fired the first step rocket, known as Bumper. It consisted of a V2 first stage and a WAC Corporal upper stage. The latter achieved a record altitude of nearly 400 km (250 miles). The last two flights of the Bumper series took place on the site that was to become one of the busiest space complexes, Cape Canaveral in Florida.

Soviet sputniks

The chief architect of the Soviet rocket program in the years following the Second World War was Sergei Pavelvich Korolev. Born at Zhitomir in the Ukraine in 1906, he developed a passion for rockets in the late 1920s. In 1931 he was a founder member in Moscow of the Group for the Study of Reactive Motion (GIRD), which can be seen as analogous to the VfR in Germany. GIRD two years later fired the Soviets' first liquid-propellant rocket. During the war Korolev was involved in designing jet-assisted take-off motors for aircraft, and afterwards witnessed the firing by the Americans of a captured V2 at Cuxhaven in October 1945.

Just as the American forces transferred von Braun's group of rocket experts and V2s to the United States, so Soviet forces transferred another group to the Soviet Union. Korolev supervised the test firing of the V2s at Kapastin Yar near Volgograd in 1947. He also made modifications to the V2 to give it extra range, which led to the design of super V2s, rather like the US Redstone, and to the first intercontinental ballistic missiles (ICBMs).

The Soviet Union successfully test fired the world's first ICBM, designated the SS-6 and codenamed Sapwood, on 27 August 1957. On 17 September, on the 100th anniversary of Tsiolkovsky's birth, Korolev gave a speech at the Palace of Unions in Moscow.

◀ **Bumper launch**
A two-stage Bumper rocket lifts off from the White Sands Proving Ground, New Mexico, in the late 1940s. But the last two of the eight Bumper launches take place at the new Long Range Proving Ground at Cape Canaveral, Florida.

He ended thus: 'In the very near future, the first experimental launchings of artificial Earth satellites will take place in the USSR and the USA for scientific purposes.' Certainly it was known that the Americans were trying to launch a satellite as part of the scientific investigations of the Earth that were currently taking place in what had been designated International Geophysical Year (IGY). But Korolev had something more specific in mind.

On 3 October he was supervising the installation of a Sapwood rocket on the launch pad at Baikonur Cosmodrome in Kazakhstan in Central Asia. Inside its nose cone was housed an aluminum sphere some 58 cm (23 inches) in diameter, with four long whip-like antennae attached. Next day he watched the rocket shoot into the sky. A little over an hour and a half later clear signals came through the radio receiver of the control room: 'Beep...Beep ...Beep...' The little sphere, called Sputnik 1, was now circling the Earth in orbit, having become the first artificial satellite of the Earth. An elated Korolev spoke to the launch crew:

'Today we have witnessed the realization of a dream nurtured by some of the finest men who ever lived, including our outstanding scientist Konstantin Eduardovich Tsiolkovsky. Tsiolkovsky brilliantly foretold that mankind would not forever remain on the Earth. The sputnik is the first confirmation of his prophecy. The conquering of space has begun.'

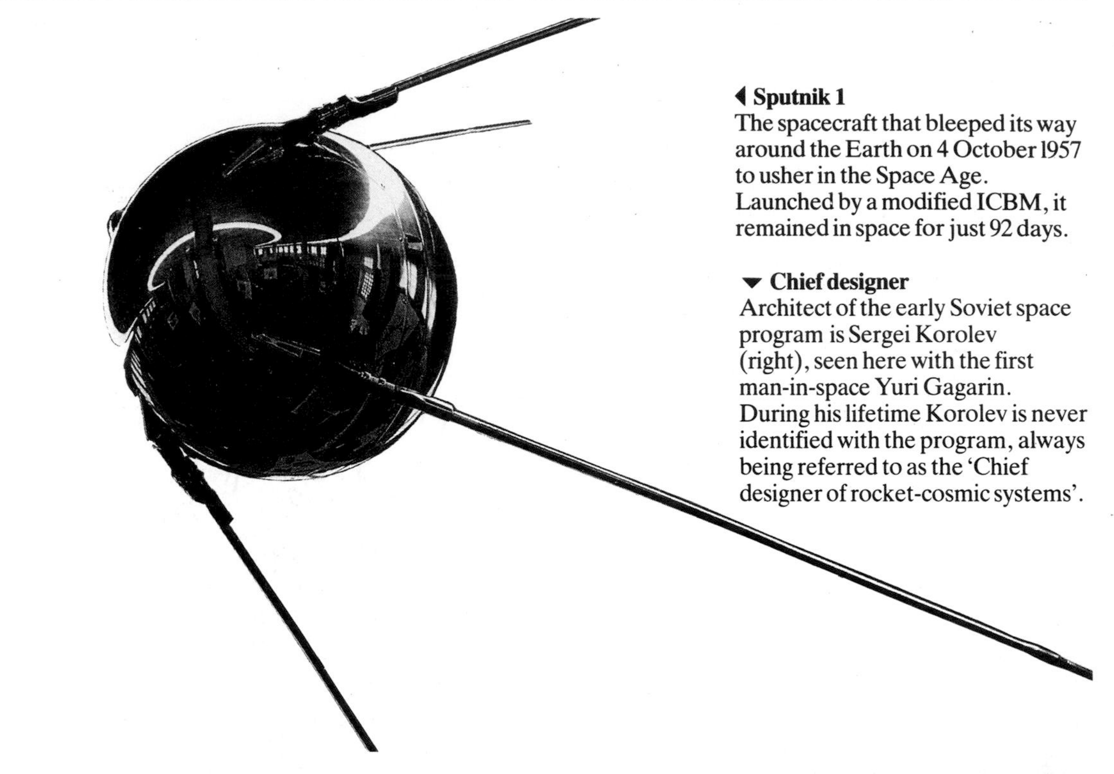

◀ **Sputnik 1**
The spacecraft that bleeped its way around the Earth on 4 October 1957 to usher in the Space Age. Launched by a modified ICBM, it remained in space for just 92 days.

▼ **Chief designer**
Architect of the early Soviet space program is Sergei Korolev (right), seen here with the first man-in-space Yuri Gagarin. During his lifetime Korolev is never identified with the program, always being referred to as the 'Chief designer of rocket-cosmic systems'.

By Jupiter

As Sputnik 1 beeped its way around an astonished world that October day, the United States renewed its efforts to get into space itself. There were two teams in contention, one from the US Navy with a Vanguard launch vehicle, the other from the Army with a Jupiter C, developed by von Braun's team. But before either was ready to attempt a launch, the Soviets launched another satellite, Sputnik 2, on 3 November. The Americans had been surprised by the size of Sputnik 1, which weighed 83 kg (184 pounds). They were absolutely stunned by the size of Sputnik 2, which weighed more than half a tonne (1120 pounds)! This was over 40 times the size of the satellite they were planning to launch! Sputnik 2 also carried the world's first space traveler, a dog named Laika ('Barker').

On 6 December Vanguard roared into life and lifted from the launch pad at Cape Canaveral aiming for orbit. But it returned two seconds later and exploded. On 31 January von Braun's team had a modified Jupiter C called Juno 1 on the launch pad, carrying an instrumented payload called Explorer 1. At 10.45 pm local time the engines ignited and this time thrust the rocket into the night sky. More than an hour and a half later signals from Explorer were picked up by the Goldstone tracking station in California, and a message flashed to the Cape: 'Explorer is in orbit.'

On 7 March the Navy at last succeeded in launching the second US satellite, Vanguard 1, which was notable by being the first satellite to be fitted with solar cells rather than batteries. In the same month President Dwight D. Eisenhower was urged to create a national agency to mastermind the US space effort, and in July he formally signed into being the National Aeronautics and Space Administration, NASA. NASA became operational on 1 October 1958 and began to shape the future.

America in space
A Jupiter C rocket streaks from the launch pad at 10.45 pm EST on 31 January 1958. One and a half hours later, Goldstone tracking station in California picks up signals from Explorer 1, the payload the rocket was carrying. The signals from the 'bird' confirms the hopes of the launch team. Explorer 1 is in orbit; America is in space!

Triumph!
Three leading members of the team that launched Explorer 1 hold up a model of it in triumph at a press conference following launch. From left to right they are William Pickering, Director of the Jet Propulsion Laboratory, which built the satellite; James Van Allen of Iowa State University, who devised Explorer's experiments; and Wernher von Braun, Director of the Army Ballistic Missile Agency (ABMA) at Huntsville, which built the launch rocket.

The age of satellites

The tentative thrusts into space of the late 1950s gave way in the 1960s to systematic exploration and exploitation of the space environment not only of near space – in the vicinity of the Earth, but also of outer space – the fathomless regions beyond. In the first year of the decade pioneering flights took place of the two classes of satellites that are now of the utmost importance to modern society – weather and communications.

Tiros 1, launched in April, was a prototype weather satellite, which tested the feasibility of using television cameras and infrared sensors to image clouds and determine temperatures. It was the precursor of a whole family of Tiros (television and infrared observation satellite) robot weather stations, the latest versions of which (designated Tiros N) are still operating.

Echo 1, launched in August, was a communications satellite (comsat), which took the form of a balloon 33 meters (100 feet) across. Radio signals were beamed up from one ground station and reflected by Echo's aluminized surface down to another ground station. The two stations were located on the East and West Coasts of the United States at Bell Telephone Laboratories, Crawford Hill, New Jersey, and at Goldstone in California. The experimental transmissions of telephone, fax and data signals were highly successful.

Echo 1 was called a passive comsat because it took no active part in relaying signals. The signals it reflected – and scattered – were very weak by the time they were received back on the ground. The answer to better reception was to equip a satellite

with on-board electronic equipment to amplify the received signals and then retransmit them to the receiving ground station. This resulted in the launch of Telstar 1 in July 1962. Test transmissions via Telstar were made over the Atlantic, not only of telephone and data signals but also of television programs. Telstar was able to relay one TV channel or 60 simultaneous telephone calls.

The 24-hour orbit

Telstar remained operational until February 1963, the same month that an attempt was made to place a comsat called Syncom 1 into geostationary orbit over the equator. In such an orbit, 35,900 km (22,300 miles) high, a satellite circles the Earth in exactly 24 hours, in the same time that the Earth itself rotates on its axis. Consequently, the satellite appears to be fixed in the sky. This obviously makes signal transmission much easier because the transmitting and receiving antennae can be permanently locked on to the satellite instead of following it across the sky as happens when the satellite is in a lower orbit. Syncom 1, however, failed on its first orbit. But its replacement, Syncom 2 (launched in July

◀ **Echo 1 in orbit**
Readily visible from the ground, Echo 1 appears in the night sky as a slowly moving star. A telescope records its passage in this long-exposure photograph.

▼ **Echo 1**
In a hanger at Weeksville, North Carolina, the first communications satellite, Echo 1, is inflated in ground tests. An aluminized balloon 33 meters (100 feet) in diameter, it is launched into orbit on 12 August 1960.

◀ **Lageos**
Launched in 1976, Lageos is a small but very heavy satellite designed to reflect laser beams. It will remain in orbit for 10 million years.

1963), climbed into geostationary orbit as planned and functioned well. It was the direct ancestor of the present generation of comsats (see page 29).

Some weather satellites, including Meteosat and GOES, also occupy geostationary orbits, but most satellites orbit much lower down. In general the lower they orbit, the shorter is their life. This is because the presence of air molecules – even in trace quantities – gradually slows them down. As they slow down, they can no longer resist the pull of gravity so well, so they fall lower. When their speed drops too low, gravity wins completely and pulls them back to Earth. When they eventually re-enter the thick part of the atmosphere, they are subjected to intense frictional heating that makes them burn up like shooting stars.

The orbit of the first ever satellite, Sputnik 1, took it at times as low as 220 km (135 miles) above the Earth. The thin traces of atmosphere at that height were sufficient to bring it down from orbit after only 92 days. In contrast the initial orbit of the first American satellite, Explorer 1, took it as high as 2500 km (1500 miles) and never lower than 350 km (220 miles). As a consequence Explorer 1 remained in orbit over 12 years, succumbing to gravity only in April 1970 after traveling 58,376 times around the Earth.

Many of the 5000 or so satellites and bits of space junk currently in orbit will remain there for centuries. One satellite, Lageos, will remain in space for literally millions of years. It is a small, very dense satellite covered with quartz prisms to reflect laser beams. It has a very precise and stable orbit and is used as a reference base for detecting movements in the Earth's crust. Because Lageos will remain in orbit for so long, it carries a pictorial message for Earth people who might find its remains when it eventually falls to Earth 10 million years hence. Carl Sagan, who devised the message, also devised that on the plaques carried by the Pioneer 10 and 11 probes, which are now heading into interplanetary space.

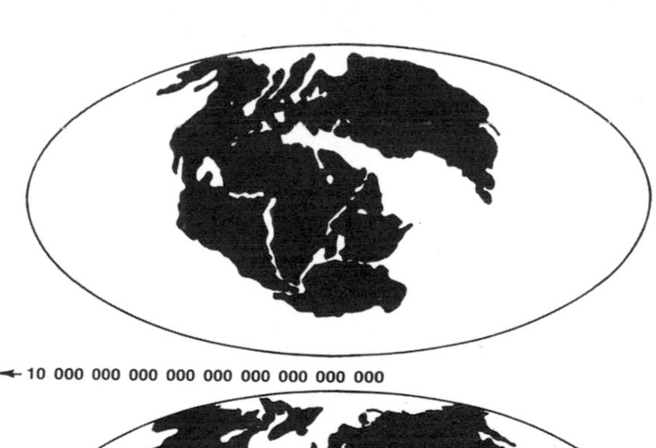

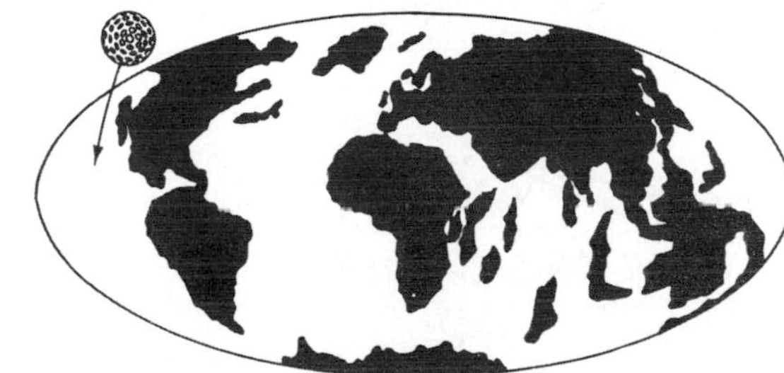

◀ (below) **Future reference**
So long will Lageos stay in orbit that it carries a coded message to inform any Earth people who may find it 10 million years hence when it was launched and whence it came. The three pictures show from top to bottom the Earth's surface as it was about 270 million years ago, as it is today, and as it will be in 10 million years, allowing for continental drift. The time scale of the three pictures is indicated beneath them in binary code.

Action and reaction

The minimum speed a body requires to get into space and remain there is about 28,000 km/h (17,500 mph). This is the speed necessary to keep it circling endlessly around the Earth, in orbit. In orbit the body is in a state of what is called free fall. It is still in the grips of the Earth's gravity and is still falling towards the Earth. However, because of its speed, the amount it falls exactly equals the amount the Earth curves away beneath it. So in fact it remains at the same height in orbit. Everything associated with the body (payload, crew and so on) is in free fall too, falling around the Earth, in a state of what is popularly termed weightlessness, or zero-g.

The speed needed to reach orbit is far in excess of any speeds experienced on Earth – it is nearly 30 times the speed of an airliner, for example. The jet engines used in airliners and even the most advanced jet fighters are just not capable of propelling something to orbital speed. They also have another major disadvantage. They are air-breathing engines and can therefore not function in airless space.

There is only one type of engine that can be made powerful enough and that can work in space – the rocket. A rocket can work in space because it carries its own supply of oxygen as well as fuel. The fuel and oxidizer (oxygen-provider) of a rocket are known as propellants. Like a jet engine, a rocket is propelled by a stream of hot gases traveling backwards out of a nozzle. It works on the principle of reaction, first stated by Isaac Newton in his famous book known as *The Principia* (1687) as his third law of motion: 'to every action, there is an equal and opposite reaction'. In the rocket, the force (action) of the gases streaming at high speed backwards out of the nozzle, sets up an equal force (reaction) in the opposite direction, and it is this that propels the rocket.

Solid and liquid propellants

Early rockets used gunpowder as a propellant – a mixture of charcoal (carbon), sulfur and saltpeter (potassium nitrate). A similar mixture is used in today's fireworks rockets. The charcoal and sulfur form the fuel. When they are set alight, they combine with the oxygen in the saltpeter to form copious amounts of sulfur dioxide and carbon dioxide gases, which form the rocket exhaust.

Solid-propellant rockets used for space launchings have a more powerful type of propellant, often containing powdered aluminum as a fuel in a kind of synthetic rubber binder. The most powerful solids used in launch vehicles are the solid-rocket boosters of the space shuttle (see page 83). Smaller solids are used as boosters for the US Delta and Europe's Ariane rockets.

A solid-rocket engine is in essence very simple, being little more than a tube filled with propellant with a nozzle at the rear end. Solids have the advantage of being cheap to build and simple to operate; they can also be stored for long periods without deteriorating. But they have the disadvantage that once they are set burning, they cannot be put out.

Most space rockets, however, have liquid propellants. Weight for weight, liquid propellants have many times the energy of solids. From the beginning, liquid oxygen (LOX) has been the main oxidizer used in rockets. Goddard used LOX, with gasoline fuel, in the world's first liquid-propellant rocket. Von Braun used LOX with alcohol as fuel in the V2. The most widely used fuel in space rockets has been kerosene. This has fueled the Russian SL series of launch vehicles, which launched Sputnik 1 and currently still launches the Soyuz ferry craft; it also fuels the four boosters of Energia. Kerosene also fueled the Atlas launch vehicles that put the first Americans in orbit and the first stage of the Saturn V Moon rocket.

Another major liquid fuel/oxidizer combination is hydrazine and nitrogen tetroxide. These propellants are termed hypergolic, which means that they ignite spontaneously when they are mixed together. The US Delta uses this combination for its second stage,

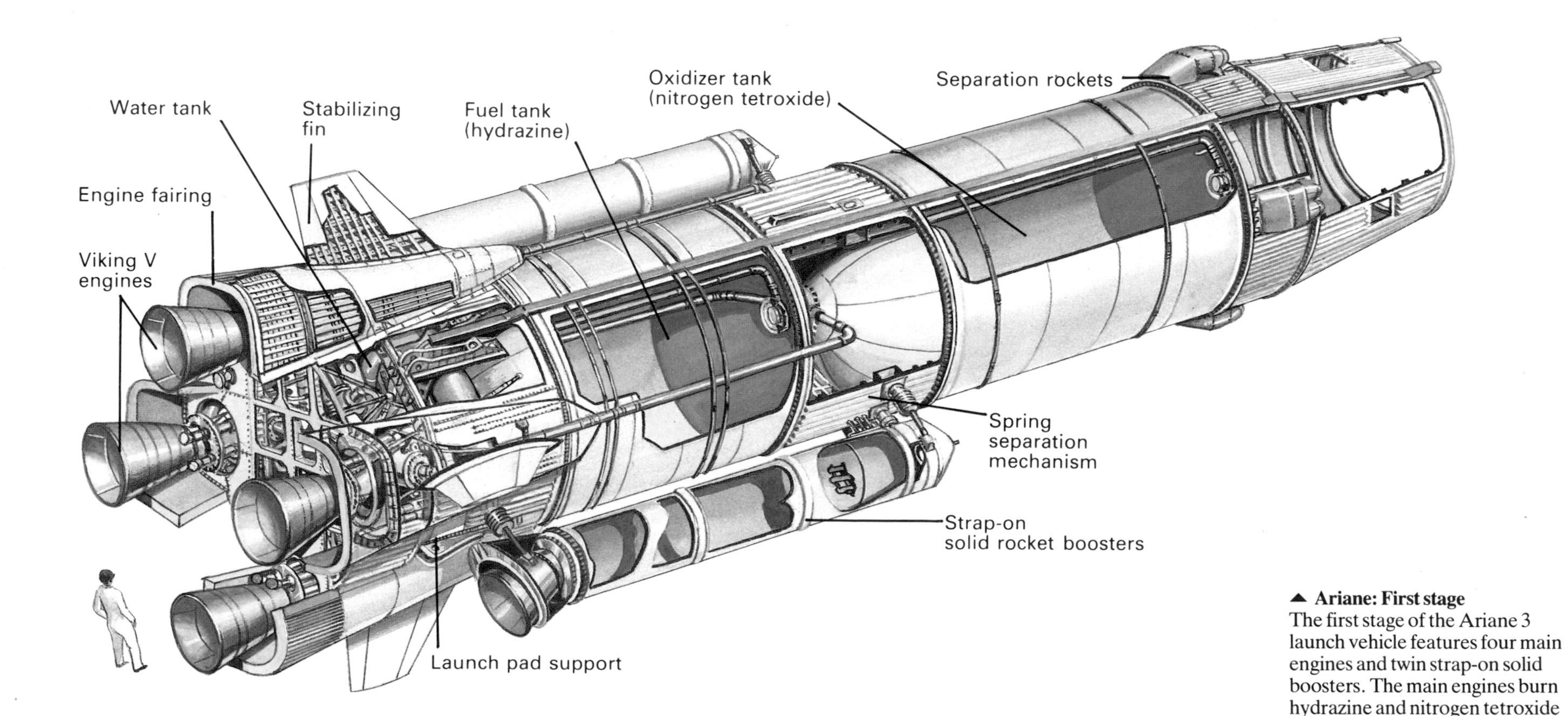

▲ **Ariane: First stage**
The first stage of the Ariane 3 launch vehicle features four main engines and twin strap-on solid boosters. The main engines burn hydrazine and nitrogen tetroxide propellants. The Ariane series of rockets are proving to be reliable workhorses.

with solid boosters and a kerosene/LOX first stage. Ariane uses hydrazine/nitrogen tetroxide for its first two stages, while China's Long March series uses it for all three, as does the US Titan.

The third major liquid-propellant combination is liquid hydrogen-liquid oxygen. This is potentially the most powerful combination of fuels there is. Ariane and the Atlas-Centaur use it for their upper stage, as did the Saturn V for its second and third stages. Energia uses it for its main core stages; the US space shuttle for its main engines. Liquid hydrogen (boiling point −253°C) and liquid oxygen (boiling point −183°C) create problems in handling because of their extremely low temperatures. They are termed cryogenic propellants, from a Greek word meaning cold.

Liquid-propellant rockets are very much more complex than solids. The fuel and oxidizer have to be stored separately in tanks,

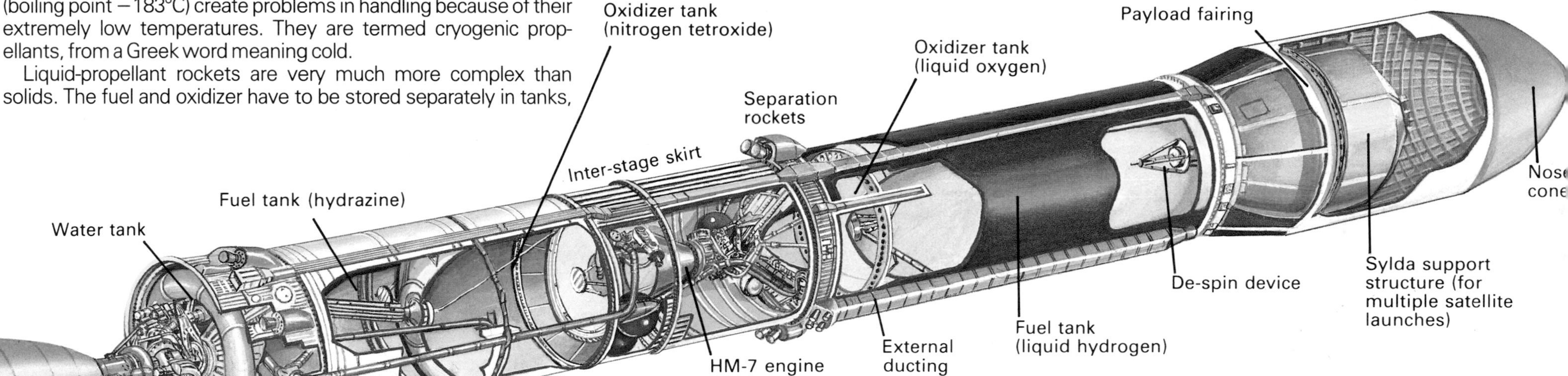

which in fact take up most of the space in the rocket. From the tanks the fuel and oxidizer are pumped into the chamber where combustion (burning) takes place. The turbopumps that do this are themselves spun by gas produced in a gas generator. The fuel and oxidizer enter the combustion chamber through an injector, which creates a very fine spray and allows fuel and oxidizer to mix intimately. To begin with, the fuel/oxidizer mixture has to be ignited (unless it is hypergolic), but thereafter will continue to burn on its own.

The hot gases produced by combustion expand rapidly out of the engine nozzle to produce thrust to propel the rocket. To extract the maximum energy from the gas stream, the nozzle is bell-shaped. Inside the nozzle, the pressures rise to over 200 times atmospheric pressure, and temperatures soar to 3000°C or more. This is high enough to melt the metal walls of the combustion chamber and nozzle. To prevent this happening, these parts have double walls, through which cold fuel is circulated on its way to the injector. This set-up – known as regenerative cooling – not only keeps the walls cool but also preheats the fuel.

▲ Ariane: Upper stages
The second and third stages of Ariane feature single engines. The second stage burns the same propellants as the first, while the third stage burns liquid hydrogen and liquid oxygen. The nose cone can accommodate as many as three satellites up to a weight of about 2.5 tonnes (for geostationary orbit).

Step by step

However powerful the propellants, no single rocket can lift itself and its payload (cargo) into orbit. The weight of the rocket structure plus propellants is always too great for the thrust it produces. There

◀ **Soviet boosters**
Unlike usual Western practice, the Soviets assemble their launch vehicles horizontally. Their vehicles are characterized by a cluster of liquid-propellant booster rockets. These rockets burn kerosene and liquid oxygen as propellants. This particular launch vehicle is being readied for a mission to Venus, Earth's closest neighbor and a favorite target for the Soviets.

▶ **GOES installation**
A GOES weather satellite, newly installed into the nose cone of a Delta rocket. As with many satellites, its body is covered with thousands of solar cells which generate electrical power from sunlight. The Delta will blast GOES into a geostationary orbit 35,900 km (22,300 miles) above the equator.

is only one solution to the problem – the multistage or step rocket. This is a launch vehicle made up of two or more (usually three) separate rockets (stages) joined together, usually end to end.

The principle behind the step rocket is this. The bottom rocket (first stage) fires first. It lifts the vehicle high into the air, and when its fuel runs out separates and falls back to the ground. The second stage fires and thrusts the vehicle, now very much lighter, higher and faster still. Then it separates in turn, and the third stage fires to boost what remains into orbit. By successively shedding unwanted weight in this way, a favorable overall power-to-weight ratio is attained that can place payloads in orbit at speeds of some 28,000 km/h (17,500 mph) or boost them even faster (to more than 40,000 km/h, 25,000 mph) so that they escape from the Earth's gravity completely to journey to the planets and beyond.

The present workhorses of the space industry, such as the Deltas, Arianes and Long Marches, are typically between about 35 and 50 meters (115 and 165 feet) long and have a lift-off thrust of up to about 600,000 kg (1.3 million pounds). Yet they can carry into orbit payloads of just a few tonnes. Over nine-tenths of their bulk is taken up by the propellants. Although it weighs nearly 200 tonnes on the launch pad, the Delta can carry less than 1.5 tonnes into geostationary orbit; Ariane 4, weighing 450 tonnes on the pad, can carry only about 4 tonnes. In other words launch rockets have an effective payload-carrying capacity of less than one-hundredth their lift-off weight. Such is the force of gravity!

Return of the giants

Dwarfing the current generation of Deltas, Arianes and Long Marches was the Saturn V, a launch vehicle specifically developed to launch men to the Moon (see page 54). It was indeed a monster. 111 meters (365 feet) long, with a take-off thrust of some 3500 tonnes (7.5 million pounds).

Saturn V made its farewell appearance on the space scene, boosting space station Skylab into orbit, in May 1973. Many observers, looking to the forthcoming reusable shuttle era, considered the giant expendable launch vehicle to be as dead as a dinosaur. Saturn Vs left over when the Apollo program was curtailed were fit only for museum pieces and tourist attractions. And they indeed ended up as such at the Kennedy Space Center at

Cape Canaveral, the Johnson Space Center at Houston and the Alabama Space and Rocket Center at Huntsville.

However, in 1987 the Soviets demonstrated that the days of the monster rocket are far from numbered. On 15 May of that year they launched what the noted Soviet space watcher James Oberg called 'the most powerful rocket in the world – ever'. Known as Energia, it made an impressive debut, lifting off from the Baikonur Cosmodrome, the Soviet's premier spaceport.

Energia, 60 meters (197 feet) high on the launch pad, is nowhere near as tall as the Saturn V, but it packs an equal if not greater punch. And whereas Saturn V was the great rocket of yesteryear, Energia is very much a rocket of today, utilizing sophisticated modern technology. On its maiden flight it carried only a dummy payload, but it is capable of lifting into orbit payloads weighing up to 100 tonnes – four times what the space shuttle can lift. As a Moscow TV commentator put it, Energia can lift into space 'the blocks from which cities will be built'. It will later be brought into operation to launch modules like the existing Soviet space station *Mir* to be assembled in orbit into an extensive space station

◀ **Delta: Launch**
Nine strap-on solid rocket boosters fire to help lift the Delta rocket off the launch pad. The Delta is one of the most reliable vehicles in America's launch stable. The satellite it is carrying here is a comsat headed for geostationary orbit 35,900 km (22,300 miles) high.

▼ **Delta: Booster jettison**
A long-range camera at Cocoa Beach snaps this fascinating picture of the nine boosters separating from a Delta rocket 87 seconds after its launch from nearby Cape Canaveral.

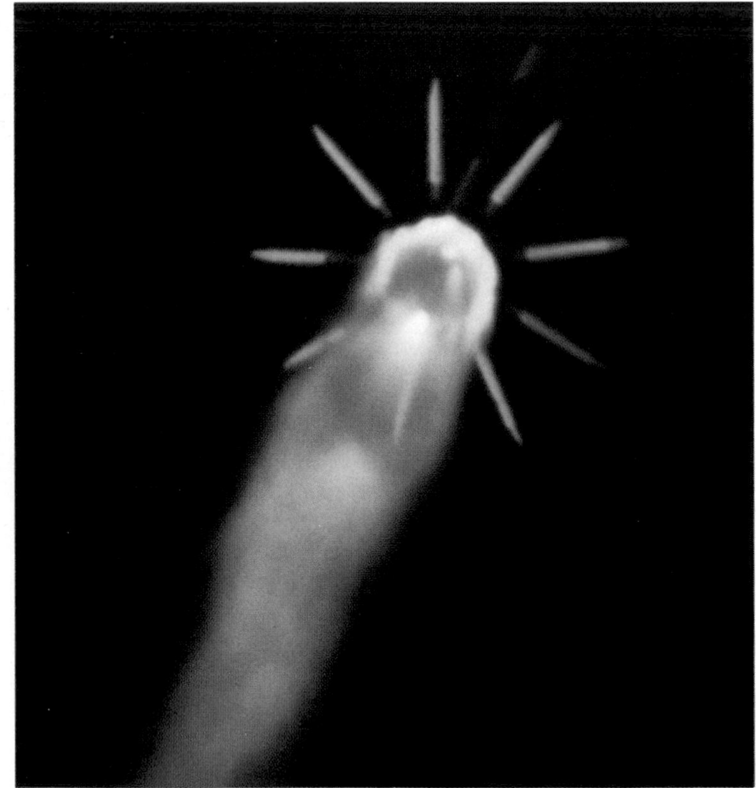

complex. Judging by past experience, the Soviets will undoubtedly attempt to do this before the international space station *Freedom* becomes a reality in the mid-to-late 1990s.

In the meantime Energia has another important role to play – launching 're-usable orbital spaceships', or Soviet space shuttles. It first demonstrated this capability on 15 November 1988, when it blasted off from Baikonur carrying an orbiter named *Buran* (*Snowstorm*). The flight was unmanned and wholly successful. *Buran* shares many features with the US shuttle orbiter, with the significant difference that it relies totally on Energia to lift it into space (see page 99).

Energia is a two-stage launch vehicle. Its four main engines, which are located in a central core stage, burn liquid hydrogen and liquid oxygen propellants. Strapped to the main core stage on the launch pad are four large boosters which, unlike the US shuttle's solid rocket boosters (SRBs), have liquid propellants – kerosene and liquid oxygen. These four boosters, so typical of Soviet rocketry, cut loose from the core stage at a height of about 50 km (30 miles). They are not recovered, falling to the ground in an uninhabited area in the eastern USSR. In turn, the core stage separates at an altitude of about 110 km (70 miles) and falls away to splash down in the Pacific. The payload module or orbiter – whichever is being carried – fires on-board engines to inject it into orbit.

Launches by shuttle

All the launch vehicles mentioned so far are expendable – in other

▸ Energia: On the pad
Ready for launch, Energia sits on the launch pad at Baikonur Cosmodrome. Note the massive main core rocket and two of the four strap-on boosters. Neither boosters nor the main core rocket are recovered.

▾ Energia: Lift-off
Packing the power of 170 million galloping horses, Energia lifts off the launch pad at Baikonur Cosmodrome spectacularly at night on 15 May 1987. On this flight it carries a dummy payload, which fails to reach orbit. A year and a half later it will launch the first Soviet shuttle.

SIEMENS

▶ **Goddard control**
America's main satellite network is under the overall control of the Goddard Space Flight Center at Greenbelt, Maryland. It receives tracking information and telemetry from stations in the worldwide satellite tracking and data network (STDN).

◀ **Earth station**
One of the European Space Agency's satellite tracking stations at Villa Franca del Castillo near Madrid, Spain. It has a 15-meter (50-foot) diameter antenna.

words, they can be used only once. Nothing comes back to Earth that is reusable, which represents a great waste of expensive hardware. And it is for this reason in the 1970s that the United States decided to switch its launch capability to a reusable system – the space shuttle (see Chapter 3).

On paper, the system promised enormous benefits in cost and accuracy of delivery to orbit. And because the system was manned, satellites could be checked out in space to ensure that they worked before they were left. However, as the shuttle era progressed, it became increasingly evident that putting all American launch eggs in the space shuttle basket was not going to be the solution to insuring routine and regular access to space. Shuttle launchings proved to be dearer than expected and could not be made as frequently as had been hoped. And they came nowhere near to meeting the demands of commercial companies – and the military – who were standing in line to place satellites into orbit.

The realization was brought into sharp focus in January 1986 when the *Challenger* disaster (page 132) called a halt to shuttle flights for 32 months. NASA immediately put out contracts to companies to supply more of the expendable launch vehicles, the Deltas and the Titans, which it had earlier phased out. And it began rescheduling its upcoming launch program for a mixed fleet of expendable launch vehicles and shuttles. Elsewhere in the space world, of course, expendables provided the only means of getting into space anyway.

Space international

For many years the Soviet Union and the United States had a virtual monopoly on space activities. But this is certainly not the case today. France was the first nation to break the US/Soviet monopoly in November 1965, when it launched its first satellite, A-1, with a Diamant rocket from Hammaguir Base in the Sahara Desert. Five years later, in February, Japan launched its first satellite, Osumi. Two months after this, China became the world's sixth spacefaring nation with the launch of Chincom 1, which distinguished itself by broadcasting music from orbit, a ditty entitled 'The East is red'!

In October 1971 Britain entered the space arena when a Black Arrow rocket launched the satellite Prospero from the rocket range at Woomera in Australia. It was India who next joined the exclusive space club in July 1980, while Israel joined as recently as September 1988, with the launch of Horizon 1.

The United States has the world's most public operational space facilities, operated by NASA. The main launch sites for satellites are at Cape Canaveral Air Force Station in Florida and, for polar launches, the Vandenberg Air Force Base in California. A few launches also take place, using the all-solids Scout launch vehicle, from the Wallops Flight Facility on the Atlantic coast of Virginia.

Operational control of US satellites is exercised by the Goddard Space Flight Center at Greenbelt, Maryland, some 16 km (10 miles) north-east of Washington DC. Goddard is also the main nerve center of the worldwide NASA communications network (NASCOM). It receives much satellite data from the White Sands Facility in New Mexico, including that from Landsat and TDRSs (tracking and data relay satellites). The main nerve center for manned space flights is the Johnson Space Center at Houston. This is the home of Mission Control, whose famous callsign is 'Houston'.

As far as numbers are concerned, the Soviet Union dominates the space scene, launching upwards of 100 satellites every year from two major launch sites or cosmodromes. The busiest is Plesetsk, a remote military facility in the far north of the Soviet Union, some 800 km. (500 miles) north of Moscow. It is used for satellites that need to go into polar orbit – over the North and South Poles. Much farther south is Baikonur Cosmodrome, the chief launch site for satellites that need to go into equatorial orbit and for manned space flights. It is located in a flat desert region west of the inland Aral Sea. The nearest town is Tyuratam. Baikonur is a sprawling facility nearly 10 times more extensive than the Kennedy Space Center and recently uprated to support Soviet shuttle launches and landings. It has no towering structures like the Vehicle Assembly Building at Kennedy spaceport, because its launch vehicles are assembled horizontally rather than vertically, and then up-ended on the launch pad.

Europe's space activities are coordinated by ESA, the European Space Agency, which came into being in 1975 from the amalgamation of two previous organizations ESRO and ELDO (European Space Research Organization and European Launcher Development Organization). Thirteen countries now belong to ESA: Austria, Belgium, Denmark, France, West Germany, Ireland, Italy, the Netherlands, Norway, Spain, Sweden, Switzerland and

▲ **Intelsat V: Preparation**
This powerful comsat, now in geostationary orbit, can handle 12,000 simultaneous telephone conversations. It relays communications signals between Earth stations on different continents by means of its directional antennae.

▶ **Mapping the Gulf Stream**
A spectacular image obtained by processing Nimbus 7 weather satellite data to identify sea temperatures. It shows in vivid mauve the warm Gulf Stream flowing to the north of Scotland.

the United Kingdom; Finland is an associate member.

ESA's launch site is located at Kourou, in French Guiana in the north-east corner of South America. From this site ESA launches its Ariane series of rockets through the company Arianespace. Kourou is an excellent launch site because it is located within a few degrees of the equator. This is ideal for payloads that have to be inserted into geostationary orbits over the equator. Also, launches can take maximum benefit from the rotational speed of the Earth. ESA has its main spacecraft testing center, ESTEC (European Space Research and Technology Center), at Noordwijk in the Netherlands. At Darmstadt in West Germany is the main communications center, ESOC (European Space Operations Center).

ESA is involved in manned space flight in collaboration with NASA. It built the Spacelab space laboratories designed to be carried by the shuttle (see page 194). It is also designing manned modules for the forthcoming space station *Freedom* through its Columbus program, together with a space plane called *Hermes*, to be launched by a powerful new rocket of the Ariane series.

Japan's growing space industry is organized by NASDA, the National Space Development Agency, in collaboration with ISAS, the Japanese Institute of Space and Aeronautical Sciences. The main Japanese launch sites are on the island of Tanegashima at the southern tip of the country and at Kagoshima just to the north.

A high point in Japan's space program was the launch of two probes, Suisei and Sakigake, to observe Halley's comet in 1986 as part of the flotilla of craft, which also included Europe's Giotto and two Vega probes from the Soviet Union. (The US contribution to this once-in-a-lifetime rendezvous with a relic of the early solar system was noticeable by its absence!) Japan is also participating in space station *Freedom*. It is supplying one of the pressurized units, known as JEM (Japanese experimental module).

China's space effort is concentrated at two launch sites. Most launches have been conducted at the Jiuquan Space Center in Gansu (Kansu) Province on the edge of the Gobi Desert in the north. In recent years a new site has been developed in Sichuan (Szechwan) Province in the south, the Xichang Satellite Launch Center, which is located at approximately the same latitude as the Kennedy Space Center (28° North). The Xichang Center is the main launch site for one of the latest versions of the Long March (Changzheng) series of launch vehicles. China is now offering its Long March 3 and 4 launchers for commercial satellite launchings on the worldwide market, and their prices are keenly competitive with those of the US Titan 3 and Europe's Ariane 4.

India's space activities are managed by ISRO, the Indian Space Research Organization. Its main design and development center is at Bangalore, while the main launch site for its all-solid satellite launch vehicles (SLVs) is the Shar Launch Center on the island of Sriharikota, north of Madras.

Satellites at work

Since the Space Age began, over 3500 satellites have been placed into orbit. At any one time only about 200-300 of them are actually functioning. The others are dead, contributing to the increasing amounts of 'junk' accumulating in space. As mentioned earlier, communications and weather satellites are the most useful satellites of all.

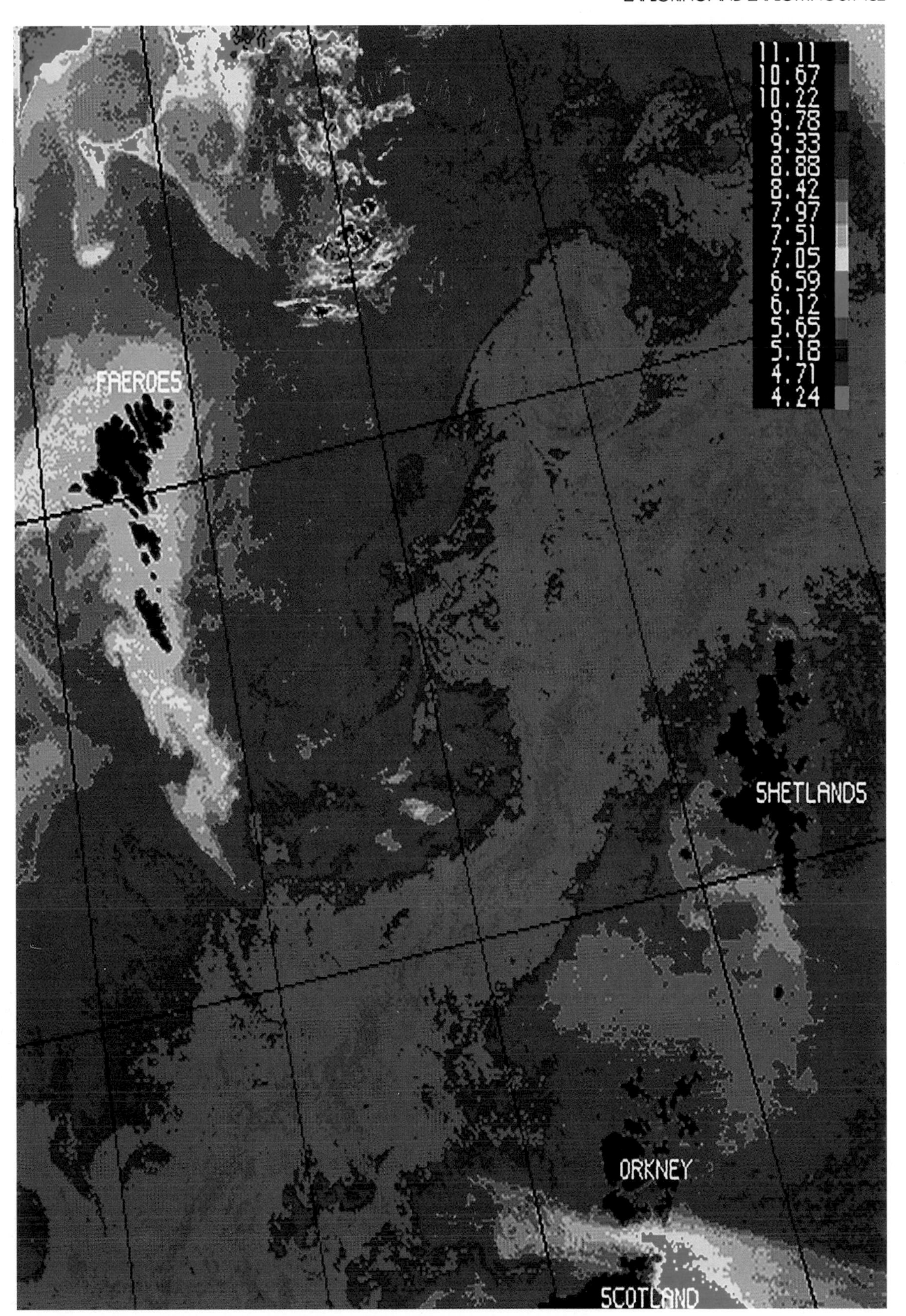

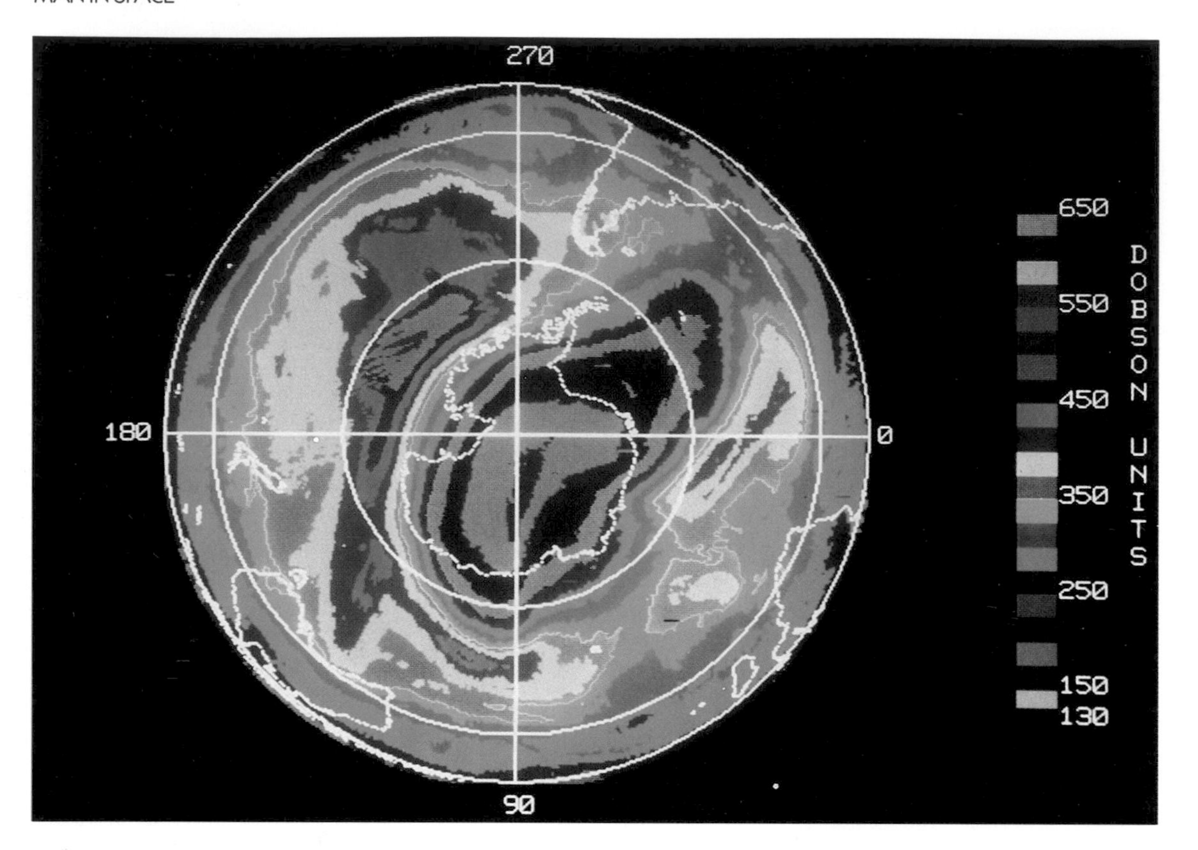

▲ **Nimbus 7: Ozone hole**
Nimbus 7 is one satellite monitoring ozone levels all over the globe. On 10 October 1986 it detects a virtual 'hole' in the ozone layer above Antarctica. It shows up in the center of this computer-generated picture as an elongated oval of light and dark violet straddling the continent. The hole is surrounded by a ring of high ozone, colored yellow, green and brown.

In the West the main international comsat network is organized by Intelsat (International Telecommunications Satellite Organization). Set up in 1964, it now has over 110 member nations. Intelsat operates groups of powerful comsats located in geostationary orbit over the Atlantic, Pacific and Indian Oceans. The latest of the series is the Intelsat 6. Of cylindrical shape, with a set of antennae on top, it measures 3.6 meters (12 feet) in diameter and 12 meters (40 feet) long. It can handle up to 30,000 telephone conversations at once.

The Soviet Union also has an extensive comsat program, which operates through the so-called Orbita communications network. Because its main population centers are located in the north, it does not use geostationary satellites like Intelsat. Instead it uses satellites called Molniya, which circle the Earth in a very eccentric, highly elliptical orbit. This takes them typically up to about 40,000 km (25,000 miles) over the Soviet Union and as low as about 600 km (375 miles) on the opposite side of the Earth. In this orbit it is 'over the horizon' for the Soviets for much of the time, although the tracking stations have to follow it across the sky.

Some comsats carry circuitry dedicated specifically to communications for shipping, enabling ships to link-up with other communications networks. This system operates through Inmarsat (International Maritime Satellite Organization). Ships also benefit from satellite navigation. The United States has set up a Global Positioning System, a network that will eventually comprise 18 Navstar satellites. They each carry an atomic clock and send out signals from which ships can pinpoint their position to within about 15 meters (50 feet). The Soviet Union is also building up a similar system using, eventually, 24 Glonass satellites.

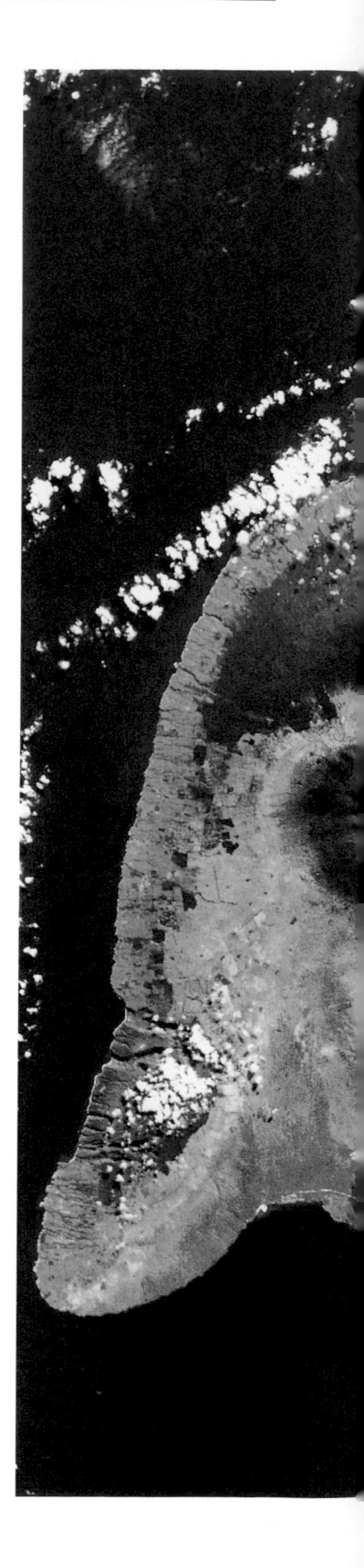

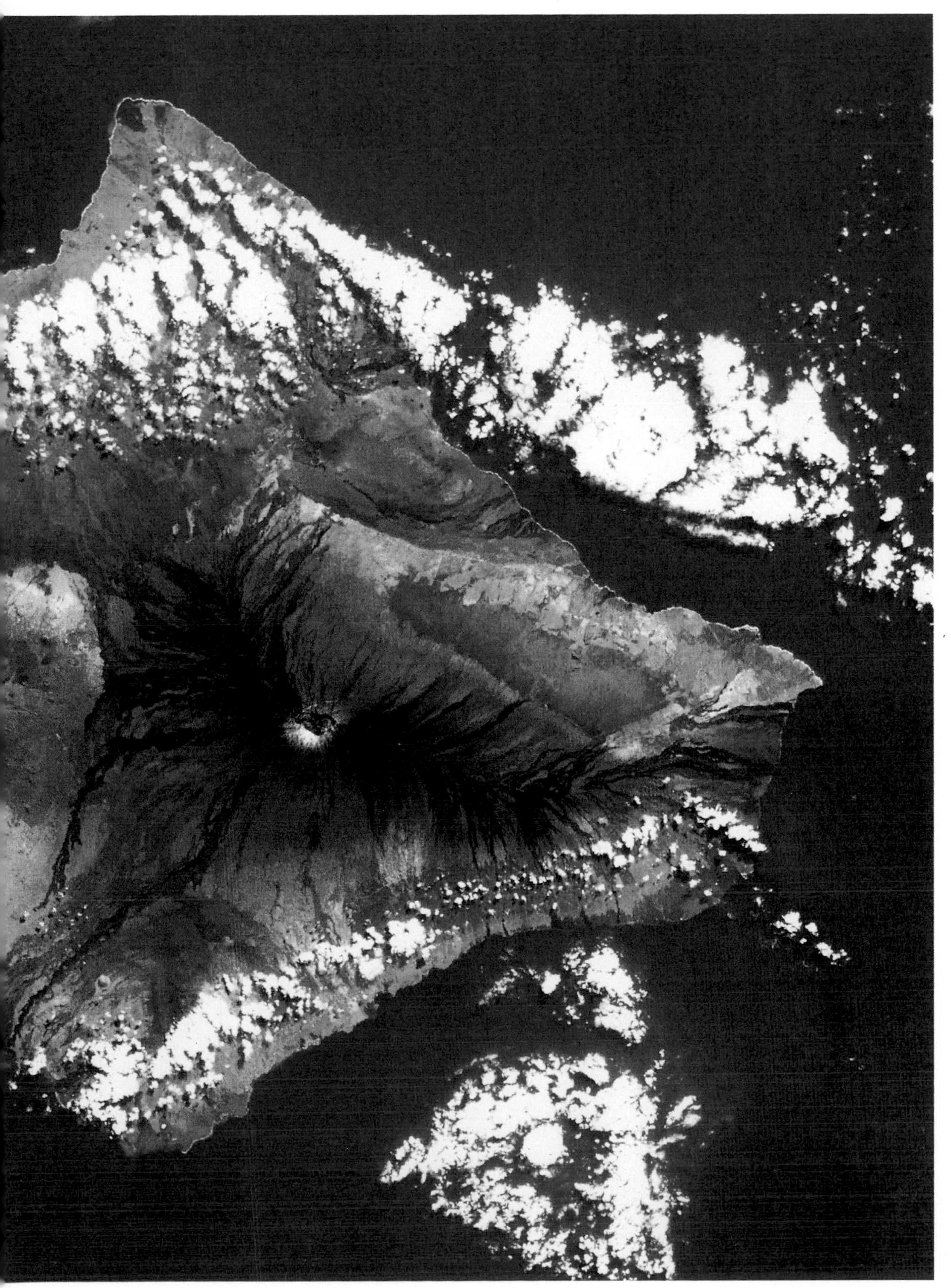

◀ **Landsat: Hawaii**
Landsat data has been processed to present the island in simulated natural color. At the center of the island is the massive volcano Mauna Loa and to the left Mauna Kea.

Weathersats

The science of weather forecasting has been revolutionized by meteorological or weather satellites, which provide pictures of cloud cover and measure temperatures of the ground, sea and at different levels in the atmosphere. Some weathersats are placed in geostationary orbit and thus remain fixed in the sky. They record the weather pattern of a whole hemisphere every half-hour. Typical are Europe's Meteosat, located over West Africa, and the US GOES, operating in two locations, over the Atlantic and the Pacific.

Many weathersats, however, are placed in a polar orbit so that they circle over the Poles while the Earth spins beneath them. In this way they can scan the entire globe once every 12 hours. Typical is the US National Oceanic and Atmospheric Administration (NOAA) series of weathersats, three of which (9, 10 and 11) are currently operational. NOAA satellites mark the latest development of the Tiros N spacecraft, first launched in 1978.

In addition to its prime role as a weather observer, NOAA carries circuitry to monitor emergency signals from downed aircraft and ships in distress and relay them to ground-sea-air rescue services. Several Cospas Soviet satellites also have this capability. The program is known as sarsat (satellite aided search and rescue). Canada and France also participate. Since 1982, when it became operational, the sarsat system has saved several hundred lives. NOAA also carries instruments to measure the concentration of ozone in the atmosphere. The destruction of the Earth's ozone layer is one of the worst environmental problems now facing us.

Remote sensing

Right from the beginning of the Space Age, views of the Earth from space have captured the imagination. Extraordinary detail can be seen from orbit even with the naked eye. During his 22-orbit trip in *Faith 7*, on the last Mercury mission (May 1963), Gordon Cooper reported seeing streets, the smoke from a locomotive and the wake of boats, from an altitude of 250 km (150 miles). Everyone back on Earth was sceptical, but photography during the Gemini missions proved that extraordinary detail can be seen from orbit. Experiments showed that in light of different wavelengths, such as infrared, hitherto invisible details become visible.

This resulted in 1972 in NASA launching the first ERTS (Earth Resources Technology Satellite), later designated Landsat 1. In turn Landsats 2 to 5 were launched, the last in 1984. Landsat 5 is the only one of the series still fully operational. It orbits about 700 km (435 miles) high, scanning the Earth in 185-km (115-mile) swaths. It scans with two instruments, the multispectral scanner and the thematic mapper. The latter scans at seven wavelengths and has a resolution of about 30 meters (98 feet). The scanners record data about reflected light electronically. The data are

Landsat: New Orleans
Lake Pontchartrain (deep blue) dominates this Landsat 4 picture. To the left lies the city of New Orleans (blue-gray) and the meandering Mississippi River (blue). The extensive red areas are regions of natural vegetation and agricultural crops. Shades of red reveal different stages of crop growth.

▶ **Solar Max**
The solar maximum mission satellite (Solar Max) is designed to monitor activity on the Sun. It malfunctions just 10 months after its launch in 1980. Four years later (April 1984) space shuttle orbiter *Challenger* on mission 41-C brings a crew to effect in-situ repairs.

▼ **Landsat: Oil slick**
By feeding information about the spectral signature of oil into the computer processing Landsat data, a false color image of the Persian Gulf is obtained which shows the presence of an extensive oil slick.

computer processed into visible images at the EROS (Earth Resources Observation System) Data Center in Sioux Falls, South Dakota.

A French Earth resources satellite called SPOT (Satellite Probatoire pour l'Observation de la Terre) currently produces the highest-resolution images. It is able to detect details as small as 10 meters (33 feet) – the size of a bus.

By careful electronic manipulation Earth-resources satellite data can yield images in a variety of false colors. The colors can be chosen to emphasize particular features of the landscape. This is possible because every type of object, such as a growing crop or an oil slick, has its own 'spectral signature', or inimitable way of reflecting light of different wavelengths. By looking for the right spectral signatures, geologists for example have been able to locate new deposits of minerals and oil. Farmers and foresters can detect disease among crops and standing timber. Hydrologists can monitor changing patterns of water resources and spot pollution. Town and country planners can more easily keep an eye on urban development and check on land use. No detail of the landscape can be hidden from a satellite's multispectral eyes.

Spysats and ASATS

Remote sensing recognizes no terrestrial boundaries, of course, and for that reason is not wholly welcomed by some countries! But for military intelligence work, greater resolution is required than that possessed by Landsat 5 or SPOT. Consequently, satellites with much more acute vision are employed. Such spy satellites are highly classified, of course, although a few details about them do leak out. The US spy satellite known as Big Bird, for example, is about 15 meters (15 feet) long, 3 meters (10 feet) in diameter and weighs about l.3 tonnes. With a powerful telescopic camera it takes very high resolution pictures on photographic film and returns this to Earth in re-entry capsules. The capsules are caught by aircraft as they drop Earthwards. The Big Birds fly in a very low orbit, about 160 km (100 miles) high, so that they have a lifetime measured only in months. They can make out objects on the ground smaller than 30 cm (1 foot). Other spy satellites are known to be able to read the numberplates of cars!

A more aggressive type of satellite is under development that is designed to attack and knock out enemy satellites. By so doing the enemy would be deprived of vital communications links and reconnaisance capability. Several types of ASATs (anti-satellite

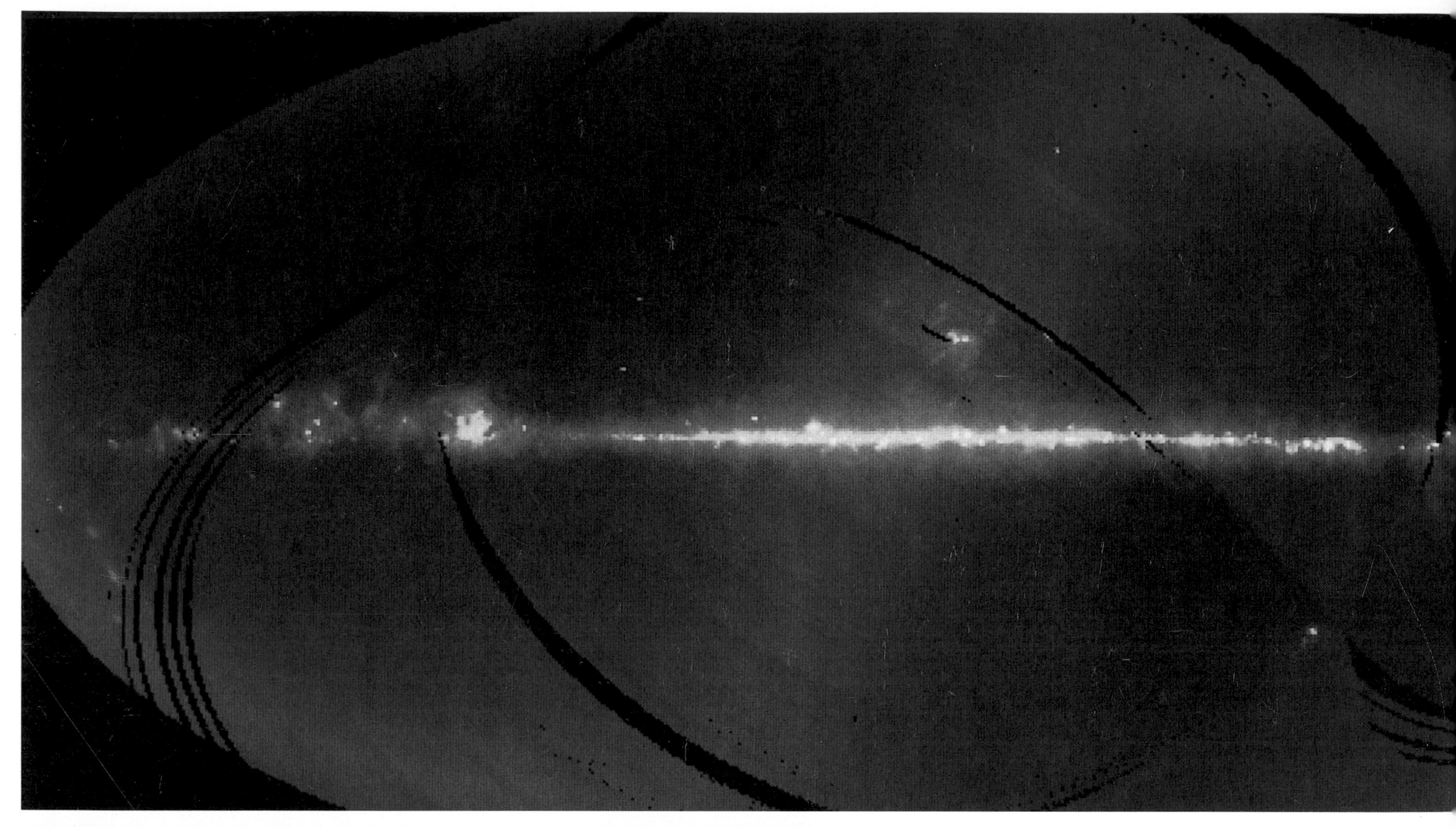

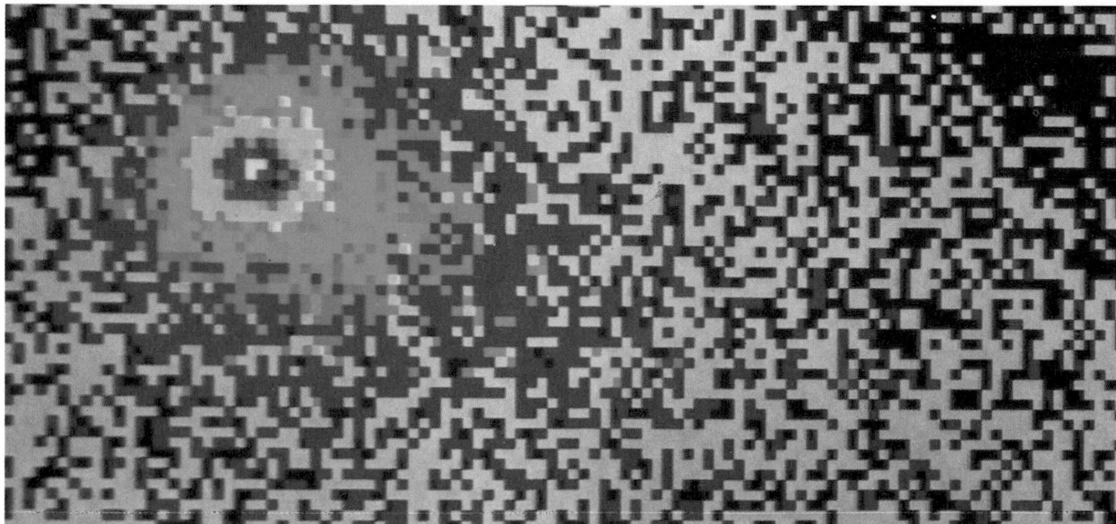

satellites) are under consideration. One type would home in on its target and then explode. The blast and shrapnel would then smash the target to pieces. Or an ASAT could contain a very powerful radio transmitter that would blow the electronic circuits of the target. Other ASATs might use lasers or particle beams to cripple enemy satellites from a distance. Space lasers and particle-beam weapons form part of the US SDI (Strategic Defense Initiative), or 'Star Wars' program initiated by President Reagan. Many people doubt whether a full-scale SDI system could operate successfully without prohibitive costs.

Probing the heavens

It was the first US satellite, Explorer 1, which pioneered space science. It carried experiments devised by James Van Allen to monitor among other things the Earth's magnetic environment. Data returned from the satellite pointed to donut-shaped 'belts' of intense radiation surrounding the Earth due to charged particles captured by the Earth's magnetic field. These were named the Van

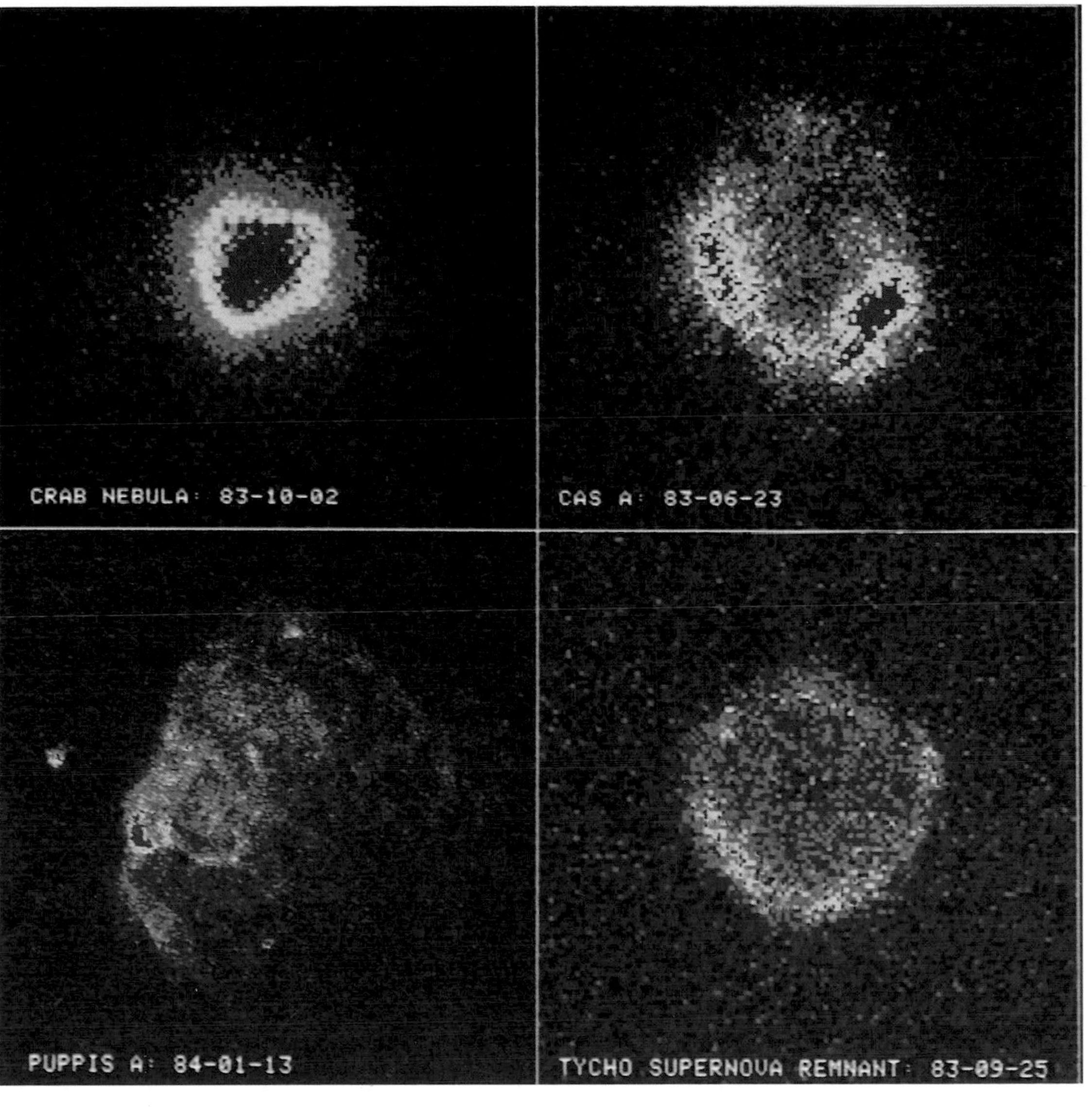

◀ **Exosat: Supernova remnants**
These four images show the X-radiation coming from four well-known supernova remnants – the remains of stars that blew themselves apart long ago. Of them, the Crab nebula is best known, originating in a supernova that was witnessed and recorded by Chinese astronomers in the year AD 1054. Cassiopeia A is also interesting. It is visually very faint, but is bright at X- ray wavelengths; and at radio wavelengths it is the brightest object in the heavens!

▲ **IRAS: Milky Way**
Data from IRAS has been combined to produce this infrared map of of our galaxy, the Milky Way. It shows most of the matter in the galaxy concentrated in the galactic plane. The isolated spot beneath the plane right of center marks the location of the Large Magellanic Cloud, the nearest neighboring galaxy to our own.

◀ **IUE: Comet Halley**
When Halley's comet returns to earthly skies in 1985/86, IUE is the first spacecraft to view it. It captures this image on the last day of 1985, when the comet's tail is beginning to form.

Allen belts.

Many subsequent satellites were launched not only to look at the near space around the Earth, but to look outwards towards the heavens. Some of these astronomy satellites looked at the universe at gamma-ray and X-ray wavelengths. We cannot detect these from the ground because they are absorbed by the atmosphere. One of the pioneering X-ray satellites was *Uhuru* (Swahili for 'freedom'), so called because it was launched from Kenya on that country's independence day in 1970. More powerful X-ray satellites, such as *Copernicus* (1972) and *Einstein* (1978) revealed quite a different universe from that we see at light wavelengths. And they found possible evidence of those strange cosmic bodies known as black holes, which have such enormous gravity that they consume everything that comes near them, even light. The European Space Agency's *Exosat*, launched in 1983, was another highly successful X-ray satellite.

ESA participated with NASA in the International Ultraviolet Explorer (IUE) astronomy satellite program, which began with the launch of IUE in 1978. Remarkably, IUE was still functioning well 10 years later, having greatly outlived its design lifespan. Again, ultraviolet radiation from the heavens is greatly absorbed, thankfully for mankind, by the Earth's ozone layer and thus can only be studied from space. Among the outstanding astronomical events of recent years covered by IUE were the return of Halley's comet in 1985 after 75 years, which was of course predicted, and the appearance in 1987 of the brightest supernova for centuries, which wasn't.

Another cooperative project by NASA, the United Kingdom and the Netherlands, was IRAS (infrared astronomy satellite). Launched in January 1983, it tuned into the infrared radiation emitted by the cosmos for 10 months. It spotted four new comets, found regions in nearby galaxies where stars are being born, and detected rings of material around the bright star Vega, which astronomers reckon could indicate the presence of a primitive solar system, perhaps not yet condensed into planets.

Chapter 2

MAN GOES INTO SPACE

◀ ***Faith 7*: Recovery**
The smile says it all. On 16 May 1963 Gordon Cooper in the capsule *Faith 7* has just been winched aboard the carrier USS *Kearsarge* after his record-breaking 22 orbits of the Earth on the final Mercury flight.

▶ **Apollo 9**
In Earth orbit in March 1969 the complete Apollo spacecraft is put through its paces for the first time, with gratifying success. Russell Schweickart snaps this picture of the CSM *Gumdrop* from the LM *Spider*.

Launching an inanimate object such as a satellite into space is one thing. Launching a living, breathing, flesh-and-blood human being is quite another. Yet this is what both American and Soviet space scientists began doing before the Space Age was four years old. Test pilots having what came to be called the 'Right Stuff' blazed the hazardous astronaut trail into orbit on the hesitant Mercury and Vostok flights of the early 1960s.

With their Gemini flights, the Americans honed to perfection the technology and techniques that paved the way for the ultimate adventure – the Apollo assault on the Moon. The Soviets, seemingly running neck and neck in a race to the Moon, faltered inexplicably at the final hurdle. And the race went to the Americans, who, against seemingly impossible odds, achieved a Moon landing by the decade's-end deadline set by President Kennedy in 1961.

The headlong rush into space took its toll of human life in both the East and the West. That there should have been casualties while pushing back the frontiers of the deadly environment of space is not surprising. The astonishing thing is that they were so few.

This chapter summarizes the early man-in-space projects up to the joint American-Soviet Apollo-Soyuz Test Project (ASTP) flight of 1975. An account of the early spacewalks is included in Chapter 5. The actual Moon landings are included in Chapter 6. The early Salyut and Skylab space stations are covered in Chapter 7.

CAUTION
FINGER TORQUE ONLY
CAUTION
FINGER TORQUE ONLY

ONCE THE TECHNOLOGY EXISTED for rocketing a satellite into space, it was inevitable that sooner or later a man would follow. The Soviets showed with their launch of Sputnik 1 on 4 October 1957 that they had a much more powerful rocket than anything the Americans had. So the odds were that they would get a man into space first. This view was reinforced by the second Sputnik launch a month later, with space dog Laika on-board in a capsule weighing nearly half a tonne.

While the panicked American military fought to get their own craft into orbit, others already had set their sights on that inevitable next step, manned flight. One was Maxime Faget, an aerodynamicist who worked at the NACA (National Advisory Committee for Aeronautics) station at Langley, Virginia. Scarcely a week after Sputnik 1's space debut, he conceived the germ of a design for a man-carrying vehicle that could survive the fiery re-entry from space. It was the birth of what evolved into the Mercury capsule, the blunt, bell-shaped craft that carried the first Americans into space.

In March 1958 at a meeting held in Los Angeles, representatives of the government, the military and the aircraft industry agreed a plan to try to beat the Soviets in the manned space flight stakes. They went for what has been termed a 'quick and dirty' solution – using existing missile rockets, such as the Redstone and Atlas, and a container, not a flying machine, to carry the man. The container became known as a capsule, and the name stuck, although the astronauts did not like it. The Air Force developed the capsule approach as part of their project MISS – man in space soonest. And it was the collaboration of the MISS team and Faget's team at Langley that evolved into Project Mercury.

On 1 October 1958 NACA ceased to exist, being transformed into a new civilian organization that would deal not only with aeronautics but astronautics – space flight – as well. That organization was NASA. Only six days after NASA opened for business, Administrator Keith Glennan announced the inception of a man-in-space program. On 26 November the program was named Mercury.

▶ The 'Original Seven'
By March 1961 the seven Mercury astronauts are in intensive training for their projected flights into space. Pilots at the summit of their profession, they frequently fly in high-speed jets like this F-102. From the left they are Scott Carpenter, Gordon Cooper, John Glenn, Virgil Grissom, Walter Schirra, Alan Shepard and Donald Slayton. All but Slayton would ride into space within the next 26 months. Slayton would not get there for 14 years.

◀ Monkeynaut Baker
In 1981, 22 years after her flight into space, Miss Baker is pictured alive and well at the Alabama Space and Rocket Center at Huntsville, Alabama.

Over Christmas 1958 President Eisenhower decided that the first men to fly into space should be military test pilots. It was the job of the newly formed Space Task Group (STG) at Langley to make the initial selection of suitable candidates from which a handful of astronauts would be chosen. Eventually 32 likely candidates were picked out, and they were then put through intensive medical and psychological testing, first at a clinic in Albuquerque in New Mexico, and then at the Aeromedical Laboratory at Wright Patterson Air Force Base in Ohio. From the 18 who emerged as 'certified supermen' (to quote *Life* magazine), a final seven were selected. They were presented to the world at large at a press conference on 2 April 1959. The 'Original Seven' were Scott Carpenter (Navy), Gordon Cooper (Air Force), John Glenn (Marines), Virgil Grissom (Air Force), Walter Schirra and Alan Shepard (Navy), and Donald Slayton (Air Force).

Spam in a can

Strictly it was not necessary for Mercury astronauts to be test pilots because the Mercury capsule was designed to fly in the automatic mode. This was termed the 'Spam in a can' approach. Such were the potential dangers of space that it was not known whether human beings could function under the excessive forces of acceleration they would experience riding into, and returning from, space. And what happened to the body in the weightlessness of orbit was anybody's guess.

However, it would not be humans that would venture into space first, but monkeys and chimpanzees. To be beaten into space by these distant relatives of man did not do much for the egos of the erstwhile astronauts! The astroprimates that ventured into space first were trained at the Holloman Aerospace Medical Center in New Mexico. They were taught to perform simple tasks inside the capsules, such as throwing levers when cued by signal lights. If they did well, they were rewarded by banana-flavored pellets, but were given an electric shock if they failed. They were quick learners!

Among the stars of the primate program were rhesus monkeys Able and Baker, which flew in a ballistic trajectory in June 1959; and Sam, which flew the following December. On 31 January 1961 a chimpanzee called Ham flew in a Mercury capsule atop a Redstone booster in a dress rehearsal for a manned suborbital flight. This flight was originally scheduled for the end of March. But it was postponed, with unfortunate consequences. For on 12 April

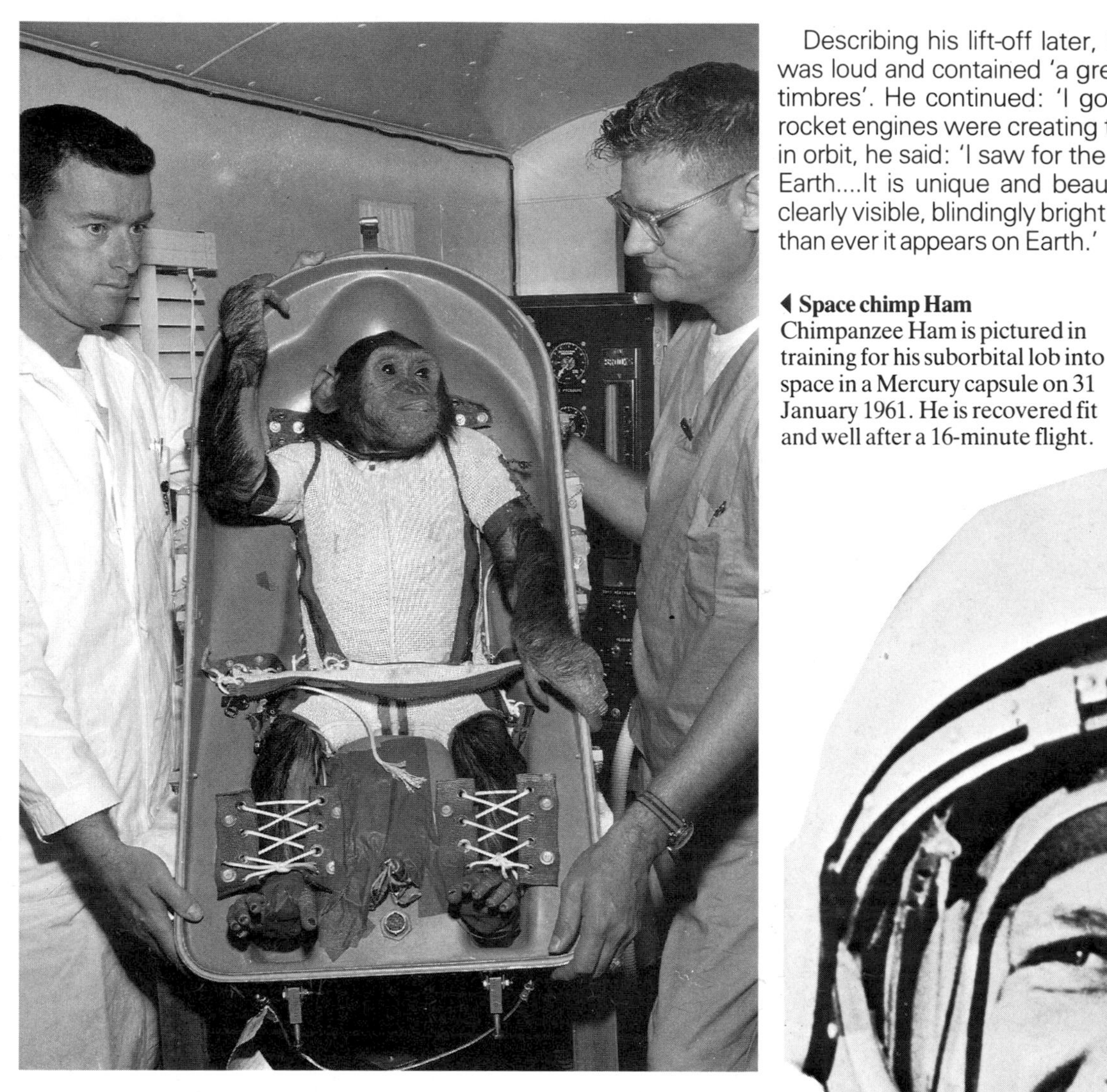

◀ **Space chimp Ham**
Chimpanzee Ham is pictured in training for his suborbital lob into space in a Mercury capsule on 31 January 1961. He is recovered fit and well after a 16-minute flight.

a Soviet Air Force Major named Yuri Gagarin rode a rocket into orbit to pioneer manned space flight. He became the world's first astronaut, or cosmonaut to use the Soviet term.

The first orbit

Just as chimpanzees preceded American astronauts into space, so dogs preceded Gagarin. As early as August 1960 the Soviets had launched two huskies, Strelka ('Little Arrow') and Belka ('Squirrel'), into orbit and recovered them alive and well. And only three weeks before Gagarin set off, another husky, Zvedochka ('Little Star'), had flown in a flawless dress-rehearsal mission.

Gagarin blasted off on 12 April at 9.07 local time from Baikonur Cosmodrome. He was riding in a spherical capsule called Vostok ('East'), nearly three times the weight of the Mercury. It took just eight minutes for the launch rocket to accelerate him from rest to a speed of 28,000 km/h (17,500 mph), many times faster than any human being had traveled before. Then the rockets cut out, and Gagarin began coasting in a circular path that took him once round the Earth. At his highest point he was nearly 330 km (206 miles) high.

Describing his lift-off later, he said that the roar of the rockets was loud and contained 'a great many new musical nuances and timbres'. He continued: 'I got the impression that the powerful rocket engines were creating the music of the future.' Of the view in orbit, he said: 'I saw for the first time the spherical shape of the Earth....It is unique and beautiful.' At night: 'The stars were so clearly visible, blindingly bright and full-bodied. The sky was blacker than ever it appears on Earth.'

▶ **Trailblazers**
The first spaceman, Yuri Gagarin (right), pictured in the summer of 1961 with the man who will shortly become the second, Gherman Titov. Whereas Gagarin only made one orbit, Titov will make no less than 17.

▼ **Spaceman No. 1**
Yuri Gagarin pioneers manned space flight when he is lofted into orbit on 12 April 1961 in a Vostok capsule. He circles the Earth once in an orbit that takes him as high as 300 km (200 miles). In his 1 hour 48 minutes flight he travels more than 40,000 km (25,000 miles).

спасибо!

UNITED
STATES
UNITED

◀ (far) ***Freedom 7*: Lift-off**
On 5 May 1961 Alan Shepard rides the Mercury capsule *Freedom 7* on a 15-minute suborbital flight. From lift-off, here at Cape Canaveral, to splashdown the flight is flawless.

◀ ***Freedom 7*: Recovery**
Some 500 km (300 miles) from the launch pad Shepard is hauled aboard a recovery helicopter and then whisked to the carrier *Champlain*. He is unharmed by his 8300 km/h (5100 mph) journey that took him fleetingly into space.

▶ **President Kennedy**
In his famous speech before Congress on 25 May 1961 the President throws at the American people the greatest challenge in their history: land a man on the Moon and bring him safely back to Earth, and do it before the decade is out.

▶ **Honoring Shepard**
In a ceremony at the White House following *Freedom 7*'s pioneering flight, President Kennedy presents astronaut Alan Shepard with NASA's Distinguished Service Medal.

Gagarin came through re-entry unscathed. At an altitude of about 8 km (5 miles), he ejected from the capsule, then parachuted uneventfully back to Earth. He touched down just 108 minutes after leaving the launch pad. The breakthrough had been made. A new epoch in space exploration had begun.

Freedom in space

It was with a sense of anticlimax that the US got ready to launch its first spaceman, Alan Shepard, three weeks later. But his mission was not to ride into orbit, like Gagarin, but to make a suborbital lob in a ballistic trajectory up to the edge of space and back. On 5 May 1961 Shepard entered his capsule, *Freedom 7*, at 5.20 am. Four hours later, he was still on the pad. By now thoroughly disgruntled, he growled at the launch crew: 'Why don't you fix your little problems and light this candle.'

At 9.30 am they did. The Redstone rocket powered Shepard through the Florida skies for three minutes, accelerating him to a speed of over 8000 km/h (5000 mph); then it cut out and fell away. Shepard continued up to an altitude of about 185 km (115 miles). At the top of his arcing trajectory he experienced weightlessness for about five minutes.

Later, as he plunged back through the atmosphere, air resistance applied its vicious brakes, subjecting his body to retardation forces of up to 12g. Then a parachute opened to lower his capsule into the ocean. He was aloft for only 15 minutes; 'Boy, what a ride!' he exclaimed. America was jubilant, but Soviet premier Nikita Khrushchev poured scorn on Shepard's achievement, dismissing it as 'a flea jump'.

'Before the decade is out'

Two days after Shepard splashed down, space enthusiast and American Vice-President Lyndon B. Johnson presented President John F. Kennedy with a report which recommended that America should aim to land an American on the Moon as an urgent space objective. The President, smarting at the growing Soviet superiority in space, needed no second bidding. And on 25 May 1961 he gave a historic speech before a joint session of Congress, in which he set the presidential seal on America's future in space and threw down a gauntlet to its people. It was a vintage Kennedy performance:

'Space is open to us now, and our eagerness to share its meaning is not governed by the efforts of others. We go into space because whatever man must undertake, free men must fully share.

'... I believe that this nation should commit itself to achieving the goal, before this decade is out, of landing a man on the Moon and returning him safely to the Earth. No single space project in this period will be more impressive to mankind, or more important for the long-range exploration of space; and none will be so difficult or expensive to accomplish. We propose to develop alternate liquid and solid fuel boosters, much larger than any now being developed, until certain which is superior. We propose additional funds for other engine developments and for unmanned explorations – explorations which are particularly important for one purpose which this nation will never overlook: the survival of the man who makes this daring flight. But in a very real sense, it will not be one man going to the Moon if we make this judgement affirmatively, it will be an entire nation. For all of us must work to put him there.

'... If we are to go only half-way, or reduce our sights in the face of difficulty, in my judgement it would be better not to go at all.

'It is a most important decision that we make as a nation. But all of you have lived through the last four years and have seen the significance of space and the adventures in space, and no one can predict with certainty what the ultimate meaning will be of mastery of space.

'I believe we should go to the Moon.'

Nothing new

The concept of a Moon-landing mission, however, was nothing new. It had first been considered seriously by the Space Task Group at Langley, which was later to form the nucleus of the new Manned Spaceflight Center at Houston, now called the Johnson Space Center.

The STG had recommended to NASA that a lunar landing should be included in an advanced space program that should follow on from the first US man-in-space program, Project Mercury. But priority should be given first to Earth-orbital activities, to the development of hardware and software for Earth-orbiting stations and craft to ferry astronauts to and fro. A voyage to the Moon was something that could take place later, in the 1970s. In July 1960 NASA Administrator Keith Glennan had announced that this vaguely defined advanced program would be named Apollo, a name first suggested by Abe Silverstein, director of NASA's Office of Space Flight Programs.

After Kennedy's speech, however, the idea of Earth-orbiting stations was shelved for the immediate future. And Apollo became specifically a Moon-landing project. NASA had to rethink its strategy for the future. It had to refine its plans for a Moon landing and select a method of achieving it. It also had still to fulfill the main objective of Project Mercury – launch a man into space and recover him safely. Then it had to bridge the gulf between the basic technologies required for that project and the advanced technologies that a flight to the Moon would demand. With this in mind, it introduced an intermediate step to the Moon, Gemini (see page 51).

▲ ***Friendship 7:* Ingress**
Aiming for obital flight, John Glenn climbs into his Mercury capsule *Friendship 7* on 20 February 1962. He had done this several weeks before but then the lift-off had been scrubbed because of bad weather. Would he fly this time?

▶ ***Friendship 7:* Lift-off**
Atop an Atlas rocket in *Friendship 7*, John Glenn soars into the heavens in a textbook launch on 20 February. He becomes the third human being to experience the strange sensation of weightlessness. He is cleared for one, two and then the full three scheduled orbits.

Aboard the *Liberty Bell*

But first things first, America had still to make it into orbit. Shepard's brief flight had been a start, now it was time to move into a higher gear. On 21 July 1961 it was Virgil Grissom's turn for a suborbital lob in a Mercury capsule. Named *Liberty Bell 7*, the capsule had a slightly modified control system and sported a new window so that Grissom could enjoy a direct (much appreciated) view of space.

Grissom's 15-minute ride itself was uneventful, though his pulse rate soared to 163 at lift-off and peaked at 171 during re-entry. It was after splashdown that the drama began. While recovery helicopters hovered overhead, Grissom asked for a few minutes to note instrument readings before they picked him up. He removed his helmet and disconnected his suit from the capsule life-support system, then he armed the explosive escape hatch. Suddenly there was a thump and the hatch blew out.

In the heavy swell, sea water started to pour in. Grissom struggled out and began to swim for his life. Water began to fill his normally buoyant suit through an open oxygen valve. He became desperate and waved frantically at the nearest helicopter. The pilot took this to mean that he was alright, and turned his attention to

▲ ***Friendship 7*: Recovery**
Just five minutes short of five hours after launch, Glenn splashes safely down. Here he is being hoisted from the recovery ship, the destroyer *Noa*, en route for USS *Randolph*, where he will complete his debriefing.

trying to hook the capsule. Eventually he succeeded, but found the waterlogged *Liberty Bell 7* was dragging him into the sea. Finally, with wheels awash he had to cut the capsule loose. Fortunately by then a second helicopter had recovered Grissom, who growled: 'My head is full of seawater.'

Despite this setback NASA decided that it was time to head for orbit. But before they could do so, the Soviets sent up another cosmonaut on 6 August 1961, not just for one orbit like Gagarin, but for one day! He was Gherman Titov, at 25 the youngest man ever to have gone into space. In his capsule, Vostok 2, he also claimed the distinction of being the first astronaut to sleep in space and to suffer from space sickness.

Nevertheless, young Titov clearly reveled in the experience as he saw the Sun rise and set 17 times. *'Dawn-1, Dawn-1.* This is *Eagle.* I feel splendid, just splendid.... Everything is fine, everything shipshape.... The Earth ... looks like a glittering sickle.... Mountains and more mountains.... They're snow-capped.... The view is simply splendid.... The peaks are in the clouds.' When the time came to return to Earth, Titov ejected from his capsule and landed by parachute, home after the longest journey in history – nearly 703,150 km (439,500 miles).

In orbit, at last

Keeping its nerve and resisting the temptation to rush a man into orbit straightaway, NASA pressed ahead on 29 November with a chimpanzee flight in a Mercury capsule atop an Atlas booster. The chimp, named Enos, was scheduled for a three-orbit mission. It went through a well-rehearsed routine, pulling the right levers as the flight progressed. But one of the levers malfunctioned and he was unjustly rewarded with an electric shock to the soles of his feet. And his flight was curtailed to two orbits.

The preliminaries were over; now was the time for the big event. John Glenn was the man chosen to attempt the first American orbital flight. After several postponements because of bad weather, his mission, in the capsule *Friendship 7,* went into the final countdown on 20 February 1962. At 9.47 am local time the Mercury/Atlas vehicle streaked into the sky, trailing a column of smoke and steam. First the escape tower separated, then the booster rocket. Just five minutes 1.4 seconds into the flight and Glenn was in space, circling the Earth in an elliptical orbit of some 160 by 260 km (100 by 160 miles).

About 45 minutes into the flight, Glenn was passing over Africa and entering his first night-time shadow. 'I still have some clouds visible below me. The sunset was beautiful,' he reported. Later, 'broadcasting in the blind making observations on night outside,' he said, 'I can identify Aries and Triangulum.... I have the Pleiades in sight out here, very clear. Picking up some of these star patterns now.'

Over Australia Glenn reported seeing lights below. These were of the city of Perth, which had turned them on full in his honor. Not

▶ **Mercury *vs* Gemini**
The sizes of the Mercury and Gemini capsules compared. The Gemini capsule is a slightly larger 'tin can' that can accommodate two astronauts. It has opening hatches through which they can go spacewalking. It has a separate equipment module (white), which is jettisoned before re-entry. The Gemini is shown docked with a target vehicle. Docking will be a major objective of Gemini missions.

▼ **Complex 14**
The site of Complex 14, long since dismantled, from where the Mercury flights were launched. The monument honors the 'Original Seven' astronauts. At left is a plaque carrying a portrait of first-American-in-orbit, John Glenn.

long afterwards he reported excitedly: 'I'm in a big mass of some very small particles that are brilliantly lit up.... I never saw anything like it!... They're coming by the capsule and they look like little stars. A whole shower of them coming by.' As he came out of the night into a new day, the 'space fireflies' disappeared.

Then the capsule began to gyrate as an attitude thruster malfunctioned, and Glenn took over manual 'fly-by-wire' control. Meanwhile, back at the Cape, Ground Control was wrestling with another problem. Telemetry indicated that the capsule's heat shield was loose. If so, it would undoubtedly break away on re-entry. Glenn, the first American in space, would become the nation's first space martyr, incinerated inside an unprotected capsule. Fortunately it was a false alarm, although no one knew this at the time. *Friendship 7* came safely through re-entry, probably the hazardous part of any mission, and when Glenn's voice ended the communications blackout the release of tension at Ground Control was palpable. 'My condition is good,' reported Glenn, 'but that was a real fireball. Boy!'

Five minutes under five hours since he left the launch pad, and after three orbits of the Earth, Glenn splashed down safely. He returned to a hero's welcome from a justly proud and joyous nation. At a ceremony in the White House Rose Garden, President Kennedy shared the elation, but saw Glenn's flight for what it was, just the beginning. 'We have a long way to go in the space race,' said Kennedy. 'But this is a new ocean, and I believe the United States must sail on it and be in a position second to none.'

◀ (below) **Time capsule**
Under this plaque on the site of Complex 14 at Cape Canaveral is a time capsule containing technical data on the Mercury flights. It is to be opened in the 25th century.

▼ **First spacewoman**
On 16 June 1963 cosmonaut Valentina Tereshkova becomes the first woman to venture into space. After her return, space will remain exclusively male territory for 19 years. Here in London, Valentina receives the Gold Medal of the British Interplanetary Society from the then President, L.R. Shepherd.

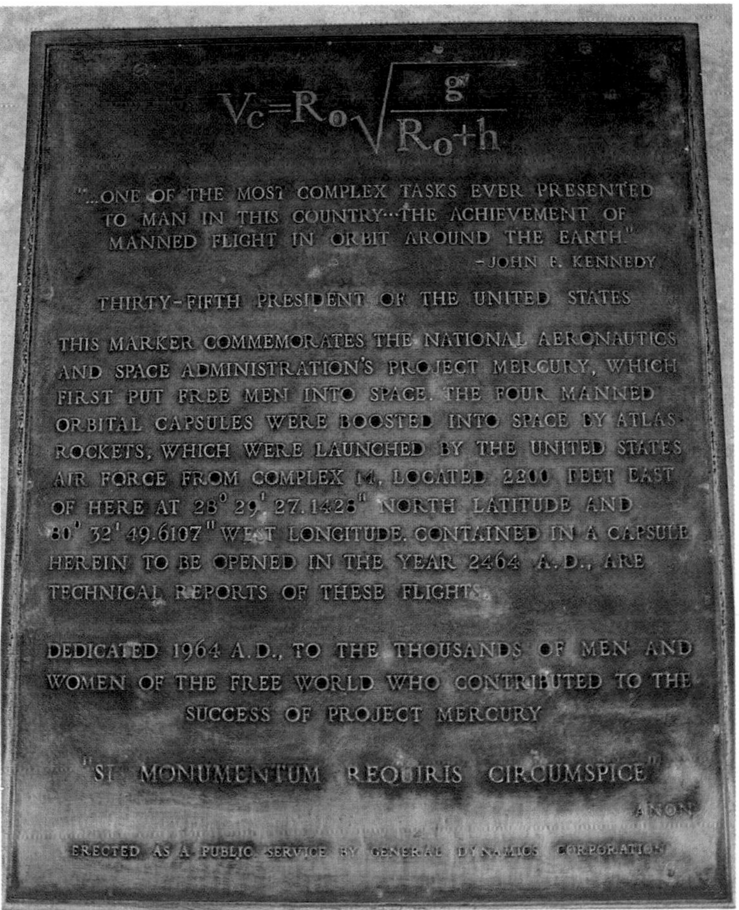

Aurora, Sigma and *Faith*

Three more flights were planned in Project Mercury before the quantum leap to Gemini. Scott Carpenter piloted the second Mercury capsule into orbit, *Aurora 7,* on 24 May 1962. It was a repeat three-orbit mission, but one that nearly went awry. Concentrating more on what he could see outside than what should have been happening inside, Carpenter overused the attitude-control thrusters, depleting them of fuel to such an extent that it threatened a hazardous re-entry. Even worse, he fired the retrorockets with the capsule at the wrong angle. All this contributed to him splashing down 400 km (250 miles) off target. He was lost to the world for 55 minutes before he was spotted by a patrol plane sitting in his liferaft. Another two hours went by before he was recovered.

On 3 October it was Walter Schirra's turn to ride the Atlas into space, in *Sigma 7,* the 'Sigma' name stressing the scientific purpose of this six-orbit flight. It was an uneventful mission in which Schirra flew mostly in automatic, chimp mode, 'drifting and dreaming,' he said.

Gordon Cooper was chosen for the final Mercury flight, which aimed for a 22-orbit stay in space. He had a particularly relaxed approach to the flight, even falling asleep on the launch pad when other astronauts' pulses would have raced! The mission began on 15 May 1963 and Cooper had more time on his hands to make observations. He startled Ground Control by the details he reported – houses in Tibet, the wake of a boat on a Burmese river and smoke from a steam locomotive. Most people found this difficult to believe and thought he must be hallucinating! Towards the end of the flight Cooper had to revert to test-pilot mode when the automatic control and stabilization system malfunctioned. He made a manual re-entry after 34 hours in space. Apart from being 3.2 kg (7 pounds) lighter because of dehydration, he was fine.

The space baby

With the flight of *Faith 7,* Project Mercury came to an end. It had achieved all its objectives. Four astronauts had spent a total of 53 hours in orbit, about 2½ man-days. While an excellent achievement in isolation, it paled by comparison with Soviet achievements up to this time. Together, Gagarin (Vostok 1), Titov (Vostok 2), Andrian Nikolayev (Vostok 3) and Pavel Popovich (Vostok 4) had notched up nearly eight days.

With two further flights in June 1963 to complete the Vostok program, the Soviets pulled even further ahead. Vostoks 5 and 6 were launched within two days of each other and came at one point as close as 5 km (3 miles). Aboard Vostok 5 was Valery Bykovsky, who went on to spend nearly five days in space. His fellow cosmonaut in orbit was the first woman in space, Valentina Tereshkova. She remained in space for nearly three days, and some accounts reported she had a pretty miserable time there. The following November she married Nikolayev (Vostok 3) and gave birth a year later to a daughter, Yelena, who was heralded as the world's first 'space baby'. Valentina's distinction of being the only woman in space lasted for 19 years, until cosmonaut Svetlana Savitskaya went into orbit in August 1982.

◀ **Gemini 12: Agena docking**
Beginning with Gemini 8, the astronauts practice docking maneuvers with unmanned Agena target vehicles. Docking will be essential if the projected Apollo missions to the Moon are to succeed.

▶ **Gemini 9: Photography**
The Gemini astronauts take breathtakingly beautiful photographs of the Earth. This one reveals most of India and (at bottom right) the island of Sri Lanka.

▼ **Gemini 6/7: Rendezvous**
On 15 December 1965 Gemini 6 homes in to rendezvous with Gemini 7 and takes this picture of it. The two craft maneuver within 30 cm (1 foot) of one another for four orbits before separating. Gemini 6 soon returns to Earth. By the time Gemini 7 returns three days later, it has set a new space duration record.

In twos and threes

In the US, after Mercury, there was a lapse of 22 months before another American went into orbit. The Americans were developing the two-man Gemini spacecraft, stepping-stone to the three-man Apollo spacecraft that would attempt to land men on the Moon. No Russian craft, either, was launched into space for 16 months after the last joint Vostok flight.

Then, on 12 October 1964, the Soviets achieved another spectacular space first by launching Voshkod 1 with a crew of three, which included a civilian scientist and a doctor. On the surface it was a considerable feat, but the main purpose was as a propaganda tool to steal the thunder from the US Gemini flights that were to begin in the following spring. Voshkod (meaning 'Sunrise') was not a new design but a stripped-down Vostok. Riding in ordinary clothes and without ejection seats, the crew undoubtedly took their lives in their hands on what some authorities reckon was one of the most hazardous space flights ever undertaken.

Even as the countdown was proceeding for the first manned Gemini mission, Gemini 3, the Soviets launched another upstaging Voshkod mission during which Alexei Leonov made the first walk in space (see page 139).

Ten for Gemini

The upstaged Gemini 3 crew waiting to get spaceborne that March were Virgil 'Gus' Grissom and John Young. Grissom had made the second suborbital lob in Project Mercury, but had yet to get into orbit. He had spent so much time with Gemini's builders McDonnell Douglas that the craft was often referred to as the Gusmobile.

The Gusmobile was obviously derived from Mercury in that the crew capsule was the familiar bell-shape. But it was somewhat longer and wider and weighed nearly twice as much. Unlike Mercury, however, Gemini was a modular design, including a so-called adapter section, which carried some spacecraft systems and consumables such as oxygen and water. At the rear of the spacecraft was a retromodule, which housed the retrorockets for the de-orbit burn. The spacecraft also had an effective reaction control system to allow maneuverability in orbit. This would permit rendezvous with other craft, a major Gemini objective. The craft also had a docking system to allow it to link up with other craft, another major objective. In addition it was fitted with opening hatches so that the astronauts could egress in orbit to go spacewalking.

Two unmanned Gemini flights in April 1964 and January 1965 preceded the first manned launch attempt on 23 March 1965. Gemini 3 was launched like all the Geminis by a Titan II modified ICBM. It was very much a test flight, lasting only for three orbits. During the brief mission Young incurred the wrath of Mission Control by offering Grissom a corned-beef sandwich! After a quick bite Grissom, known to hate the dehydrated space food, hastily secreted the offending item lest it let loose a shower of crumbs into Gemini's delicate instrumentation.

Grissom had unofficially named the spacecraft *Molly Brown*, after the heroine of a Broadway musical, *The Unsinkable Molly Brown*. The 'unsinkable' part alluded to his suborbital space flight, which ended with his capsule sinking. Grissom did not escape problems at splashdown this time either. While waiting for recovery he was seasick. 'Gemini may be a good spacecraft,' he grumbled, 'but she is a lousy ship.'

Gemini 4's flight, which began on 3 June 1965, was also a three-orbit mission, chiefly memorable for Edward White's spectacular spacewalk (see page 139). By constrast, Gemini 5 on 21 August began an eight-day endurance mission to test Gemini systems at length. In so doing it doubled the previous space endurance record set by Vostok 5 two years before. The flight also marked the first time fuel cells were used on a spacecraft to provide electrical power.

Gemini 7, launched 11 days before Gemini 6 on 4 December 1965, went in for a truly marathon flight. It did not return to Earth for little short of two weeks. Its two astronauts, Frank Borman and James Lovell, set up a space endurance record that would stand until the flight of Skylab 2 in 1973. The other outstanding achievement of the flight was the rendezvous in orbit with Gemini 6. The two craft maneuvered to within 30 cm (1 foot) of each other.

And so the procession of Gemini flights continued, new records being set with each one. From Gemini 8, the astronauts began to practice docking with unmanned Agena vehicles. They began to make longer and longer spacewalks, although these were seldom trouble-free. A variety of experiments was undertaken, and the art of orbital photography was pioneered, with breathtaking results.

The end for Gemini

The outstanding success of the first nine Gemini flights was marred by one thing only, the problems that had been generated by spacewalking. So on the final Gemini mission the scheduled extravehicular activities (EVAs) were planned down to the minutest detail. That mission began on 12 November 1966, crewed by

◀ Gemini 4: Splashdown
Minutes after splashdown of the Gemini capsule, divers are on hand to secure a flotation collar around it. Then the hatches can be opened to allow egress of the crew into the life-raft. Note the green dye used as a marker. This picture shows well the ablative heat shield on the capsule's base.

▼ Apollo: Heading Moonward
Having just been blasted out of Earth orbit into a translunar trajectory. The crew in the CSM perform docking maneuvers to extract the lunar module from the third stage. In just three days they will be rendezvousing with the Moon.

James Lovell making his second Gemini flight, and Edwin 'Buzz' Aldrin. As these two rode the pad elevator to the capsule level, they carried appropriate signs on their backs, 'The' and 'End'. As on previous missions, an Agena target vehicle was launched into orbit for them to dock with. However, the radar that they should have used to rendezvous with the Agena inexplicably failed. Aldrin, however, came to the rescue. Using emergency charts and sextant measurements, he guided Lovell to the target. Coincidentally, he had made orbital rendezvous his specialty in his doctoral thesis at MIT and had studied the problem they now faced in some depth. Not for nothing was he nicknamed 'Dr Rendezvous'!

After a successful rendezvous and docking, the crew maneuvered their craft to witness and photograph that most awe-inspiring of natural phenomena, a total eclipse of the Sun. They caught it over South America 'right on the money', as they reported. Over the next three days Aldrin, carefully pacing himself, performed a total of 5½ hours of EVA with conspicuous success. When he closed the hatch for the last time on day three of the mission, little did he suspect that the next time he would go spacewalking would be kangaroo-hopping on the Moon (see page 165).

Gemini 12 splashed down in triumph on 15 November 1966 and Gemini, the afterthought of a project, was at an end. During their 10 missions, 604 revolutions of the Earth and 1900 plus hours aloft,

◀ Von Braun: Rocket man
Inside the blockhouse, Wernher von Braun observes the lift-off of a Saturn IB launch vehicle (SA-203) on 5 July 1966. It puts into orbit the SIVB, the rocket that will become the third stage of the Saturn V Moon rocket.

▼ Saturn V's stages
The enormous size of von Braun's Moon rocket can be judged from this picture of one that didn't make it to the Moon. It is located at the Johnson Space Center, Houston, home of Mission Control. The launch vehicle is separated into its various stages.

the 16 Gemini astronauts accomplished all they set out to do, perfecting rendezvous and docking and, finally, overcoming the spacewalk problem. The hesitant steps into space that had marked Project Mercury had given way to confident strides. The Americans had man-rated themselves for a future in space.

Recces to the Moon

Strangely, during the 20 months the $1.3 billion Gemini program was running, no Soviet manned launches were made. And America nosed ahead in the space race for the very first time. The main reason for this undoubtedly was the untimely death of Soviet chief designer Sergei Korolev on 14 January 1966. He had been the main architect of, and provided the vigor for, the Soviet thrust into space. Without his genius and drive, the Soviet initiative began to falter.

However, observers in the West suspected that the lull in manned Soviet launchings might signal the build-up to an attempt at a Moon landing before Apollo could get off the ground. Their suspicions seemed to be well founded when, on the last day of January 1966, the Soviets launched a Moon probe, Luna 9, which landed on the Moon's Ocean of Storms (Oceanus Procellarum) and transmitted photographs of the landscape. Radio Moscow proudly called the landing 'a major step towards a manned landing on the Moon and other planets'.

Exactly three months later Luna 10 was launched and became the first probe to go into lunar orbit. Three further probes – two orbiters and a lander – were launched to reconnoiter the Moon before the year's end. By this time parallel lunar reconnaissance by the US was underway, with successful launchings of the Surveyor 1 lander in May and Lunar Orbiters I and II in August and November. So both nations appeared to be running neck and neck in a race to our nearest heavenly neighbor. Who would win?

While the world guessed at Soviet intentions, the US by the end of 1966 had decided on a date for its first manned Apollo flight, 21 February 1967. The prime crew for this mission, designated Apollo 1, were in the final stages of their intense training. The multifarious pieces of hardware and the extensive new facilities Apollo demanded were in an advanced state of readiness. Everything seemed on schedule to meet Kennedy's decade's-end deadline three years hence.

Modus operandi

It is apt at this juncture to outline the means by which Apollo sought to bring about a lunar landing and review the hardware and facilities that would be required.

When Apollo was originally conceived, three possible methods were put forward that could achieve a lunar landing. One was direct ascent. A huge rocket would blast the astronauts directly to the Moon from the Earth, and then blast them directly back again; there would be no intermediate maneuvers on either leg of the journey. This was an attractive proposition because of its simplicity, but the practical difficulties were daunting. A massive new rocket would be required (Nova), whose development would not be feasible, given the existing state of technology, within the time frame imposed by Kennedy.

Both of the other methods – Earth orbit rendezvous (EOR) and lunar orbit rendezvous (LOR) – required the development of a more modest launch vehicle and of modular craft with a rendezvous and docking capability. EOR would require the launch of two rockets, one carrying fuel, the other the spacecraft and crew. The two would first rendezvous in Earth orbit. The fuel would be transferred to the manned rocket, which would then proceed to the Moon. The rocket would return from the Moon directly to Earth.

The third method, LOR, called for a single rocket launch of a three-module spacecraft, which included a lunar lander. The lander would separate from the mother ship in lunar orbit and descend to the lunar surface. Later the lander would lift off the Moon and return to lunar orbit to rendezvous and dock with the mother ship,

◀ **Saturn V engine**
The size of the components that make up the Saturn V is huge. Here at Kennedy Spaceport's Visitors Center, the author is pictured next to an F-1 engine of the type used in the rocket's first stage. In the background is the business end of a Saturn IB, the vehicle used for flight-testing Saturn V hardware.

▲ **Command module**
The size of the command module is well illustrated here at the Johnson Space Center, where the Apollo 17 spacecraft resides. The base bears the scorch marks of its fiery re-entry through the atmosphere.

▲ (right) **Apollo: Re-entry**
The Apollo command module slams into the atmosphere traveling at a speed of nearly 40,000 km/h (25,000 mph). Its heat shield melts and begins to burn away, but in so doing it prevents the whole craft burning up.

which would then return home. LOR was the brainchild of John Houbolt, who headed a NASA study group on space rendezvous. It offered, said Houbolt, 'a chain reaction of simplifications: development, testing, manufacture, launch and flight operations'.

Despite initial opposition from EOR proponents, NASA eventually chose LOR and announced its selection officially in July 1962. One of the last to be convinced of the practicality of LOR was Wernher von Braun, whose task it would be at the Marshall Space Flight Center at Huntsville, Alabama, to develop the necessary rocketry for whatever method was selected. Indeed, his team at Marshall had already (in October 1961) launched the first Saturn I, first member of the Saturn family of heavy launch vehicles, which would lead, step by step, to the development of the Moon rocket itself, the Saturn V.

The Apollo Moonship

The three-module spacecraft which LOR demanded comprised the command module, the service module and the lunar module. The command module would be home for the crew of three. It was a cone-shaped unit some 3.5 meters (12 feet) high with a base diameter of 3.9 meters (13 feet). It was a double-walled structure with an inner pressure shell, in which the crew breathed a pure oxygen atmosphere at about one-third of an atmosphere's pressure. Outside, the most critical component was the heat shield. This was designed to withstand the 3000°C temperatures the command module would experience during re-entry into the atmosphere on return from the Moon. It was made of a phenolic/epoxy resin reinforced with wiremesh. It did its job by melting and boiling away, a process known as ablation.

For most of the journey to and from the Moon, the command module would be mated with the service module, the combination being known as the CSM (command and service modules). The service module supplied the command module with life-supporting oxygen, power and water – the power and water both coming from fuel cells. It carried the liquid hydrogen and liquid oxygen to fuel these cells, together with propellant tanks for its engine. On the outside were four sets of 'quad thrusters' – little jets facing in four directions – which provided attitude control for the CSM.

The third element in the Apollo spacecraft, the lunar module, was an odd-looking craft, consisting of two parts, an ascent stage (upper) and a descent stage (lower). Its eccentric shape was dictated purely by its function. Of relatively flimsy construction, it was designed for use exclusively under the low lunar gravity, which is only one-sixth that on Earth.

On the outward journey to the Moon, the lunar module would be mated with the command module. In lunar orbit, two of the astronauts would ride the lunar module down to the Moon's surface. After their exploration, they would return to orbit in the upper stage of the lunar module, using the lower stage as a launch pad. In orbit they would rendezvous with the CSM, still orbiting above. When the crew were reunited, they would jettison the lunar module and fire the service module's engine to boost them out of lunar orbit into a trajectory that would take them back to Earth. Just before they re-entered the Earth's atmosphere, they would jettison the service module. Slowed first by atmospheric drag and then by parachutes, they would splashdown inside the command module at sea.

The mighty Moon rocket

Altogether some 3 million parts went into the three modules that made up the Apollo spacecraft. The total weight of the craft was

◀ **Complex 39**
Dominating this complex at the Kennedy Space Center, constructed to support the Apollo assault on the Moon, is the towering VAB, 160 meters (525 feet) tall. Its prime function is to put together the Saturn V, seen at right leaving for the launch pad on its mobile launch platform. The low building to the left is the Launch Control Center.

▼ **Cosmonaut martyr**
Vladimir Komarov meets his death returning to Earth on 24 April 1967, after test-flying the new Soyuz spacecraft. Here at his funeral his widow grieves at the Kremlin wall, in which his ashes have been interred.

over 45 tonnes. To launch such a weight to the Moon required a mighty rocket of truly gigantic proportions. This rocket was the Saturn V. It was the last in the Saturn series of rockets developed by Wernher von Braun's team at the Marshall Space Flight Center. It outpowered the previous rocket in the series, the Saturn IB, by a ratio of five to one, boasting a take-off thrust of some 3.4 million kg (7.5 million pounds).

With the Apollo spacecraft and its escape rocket on top, the Saturn V stood no less than 111 meters (365 feet) tall. Its five huge first-stage F1 engines burned kerosene and liquid oxygen propellants. The five J2 engines of the second stage and the single J2 engine of the third, upper stage burned high-energy liquid hydrogen and liquid oxygen propellants. The third stage went into orbit with the Apollo spacecraft and was fired again to boost Apollo into a trajectory that would carry it to the Moon.

The Saturn V, like the Apollo spacecraft, was made up of some three million individual parts. Fully fueled on the launch pad, it weighed a colossal 2900 tonnes. To assemble and transport such a monster demanded structures, facilities and vehicles of unprecedented dimensions at the launching site on Merritt Island, a few kilometers away from the Cape Canaveral launch sites of earlier missions. The Apollo launch site, known as Complex 39, became (and is still) the heart of the Kennedy Space Center.

The focal point of the new launch complex was the building in

▶ **Apollo 1: Training**
Inside the Apollo Mission Simulator in late 1966 the prime crew train for the first manned Apollo flight. From the left they are Edward White, who made the first spacewalk for the US from Gemini 4; rookie Roger Chaffee; and Virgil Grissom, who made the second suborbital flight in Project Mercury and also took part in the maiden voyage of Gemini.

▼ **Apollo 1: Inferno!**
The charred interior of the spacecraft after the flash fire on 27 January 1967, which in seconds killed the crew and put the Moon-landing program in jeopardy. Edward White, Roger Chaffee and Virgil Grissom were just completing a simulated countdown atop the launch vehicle on the pad, three weeks away from their scheduled launch date.

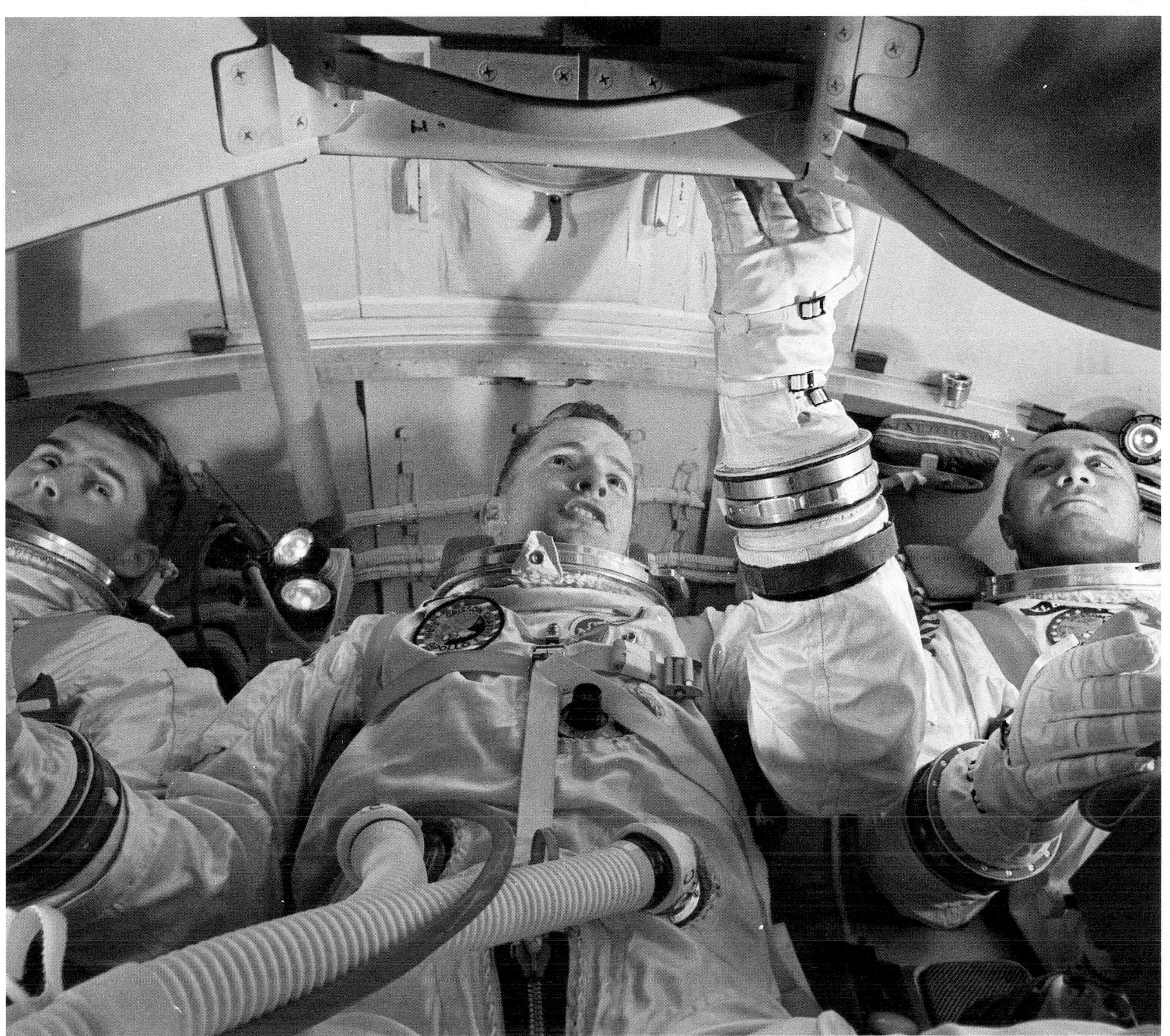

Apollo 7: Target
On 11 October 1968 the Apollo 7 mission lifts off; the objective is to man-rate the Apollo CSM, much modified since the Apollo 1 disaster. The crew practice rendezvous and simulated docking maneuvers with the upper stage (SIVB) of their launch vehicle.

Apollo 7: Recovery
After a highly successful 11-day flight, the crew of (from the left) Walter Schirra, Donn Eisele and Walter Cunningham are picked up on 22 October 1968. Walter Schirra has just completed his record third flight. At the age of 45, he is the oldest astronaut yet to venture into space.

(opposite) **Apollo 8: Lift-off**
Carrying a human crew for the first time, the Saturn V lifts itself, slowly at first, off the launch pad in the early morning of 21 December 1968. In the sky, looking deceptively close, is its target, the Moon. Within hours, the Apollo 8 spacecraft is heading out of Earth orbit and committed to a voyage into the unknown.

which the Saturn launch vehicles were put together. This Vehicle Assembly Building (VAB) ended up as one of the biggest industrial buildings in the world – a colossal 218 meters (716 feet) long, 158 meters (518 feet) wide and 160 meters (526 feet) high. It covers an area of over 3 hectares (nearly 8 acres). Its four high bays were designed so that work could proceed simultaneously on four Saturn Vs!

Inside the VAB the Saturn/Apollo launch vehicle was built up on a mobile launch platform. This carried a steel tower which provided access to the rocket on the launch pad. To get to the launch pad, some 5.5 km (3.5 miles) away, required the services of a custom-built vehicle, a massive crawler transporter propelled by four twin-crawler tracked trucks. Two such vehicles were built. They were, and are still, the world's biggest land vehicles. (Refurbished, they now carry shuttles to the launch pad.) They measure some 40 meters (130 feet) long and 35 meters (115 feet) wide. By themselves they weigh some 2700 tonnes. The combined weight of transporter, mobile launch platform and Saturn V launch vehicle was an incredible 8100 tonnes! It therefore comes as no surprise that this unique form of transportation had a top speed of only 1.6 km/h (1 mph). Well was it nicknamed the 'mighty tortoise'!

Work began on the Complex 39 site in November 1962. The first rollout of a dummy Saturn V/Apollo stack came on 25 May 1966. It was an appropriate date, five years to the day after President Kennedy's historic speech. As technicians, government representatives and astronauts waited outside the VAB for the gigantic assembly to emerge, Deputy NASA Administrator Robert Seamans remarked laconically: 'We are now going to see if the idea works!'

Countdown to disaster

Exactly three months after that first rollout, a successful third flight of the Saturn IB rocket took place, making it 13 successful Saturn flights in a row. And in October NASA announced the crew of the first manned mission which would take place on 21 February 1967. The mission was designated Apollo-Saturn 204, later to be renamed Apollo 1. The crew selected were: Virgil Grissom, who had flown the second suborbital flight in the Mercury capsule *Liberty Bell 7*, and the first Gemini flight, Gemini 3; Edward White, who from Gemini 4 had made the first American spacewalk; and rookie Roger Chaffee.

As 1967 dawned, everything was going well – the tests, the training were all on schedule for the February launch date. On Friday 27 January the crew prepared for the last major test prior to launch – a simulated countdown in their Apollo flight capsule, a Block 1 (Earth orbital) command module, *in situ* atop the Saturn IB rocket (unfueled) on the pad at Launch Complex 34. It was to be a 'plugs-out' test, which meant that the umbilicals to the rocket and spacecraft on the pad would be disconnected, as they would in a real launch.

The crew entered Apollo at about 1 pm. As their suits were looped into the pure oxygen atmosphere they would be breathing during the test, Grissom reported a strange odor in the system: 'a sour smell somewhat like buttermilk,' he said. After tests on the system, the countdown went ahead. But it was not a smooth one. Troublesome faults again reappeared in the oxygen system. A higher than normal oxygen flow, believed to be caused by movements made by the crew, periodically tripped the master alarm. There were also problems in communications between Apollo and launch control.

'I smell fire'

At 6.31 pm, as the crew and launch team prepared to come out of yet another glitch-induced hold, instruments indicated another surge in the oxygen flow to the suits. The pressure inside the capsule now registered 16.7 pounds per square inch (psi), two pounds above atmospheric pressure. It was pure oxygen. Suddenly, almost casually, Chaffee (probably) reported: 'Fire, I smell fire.' Seconds later, White cried out: 'Fire in the cockpit!'

The launch control saw on their television monitors, which were trained on the cockpit, frantic movements of arms and legs and then a burst of flame. A desperate cry rang out over the intercom, followed by an ominous silence. As stunned technicians in the white room alongside the spacecraft rushed vainly to attempt an impossible rescue, the spacecraft ruptured under the pressure built up inside. Flame and dense black smoke billowed out. There was danger now that the conflagration would cause the

emergency escape rocket atop the command module to ignite. And if that happened the whole launch complex could go up. Many of the technicians fled, but a half-dozen remained and wrestled for nearly six minutes with the bolts that secured the hatch before they were able to reach the crew. However, the astronauts were long since dead, killed by inhalation of carbon monoxide, with burns a contributory cause.

Was it with some prescience of impending doom that Grissom had said just weeks before: 'If we die, we want people to accept it. We're in a risky business.... The conquest of space is worth the risk of life.'

A minor malfunction

As investigators sifted through the charred remains of the Apollo capsule, a stunned nation, perhaps over-used to hearing only good news about the space program, began to query the whole ethos of space flight. Perhaps after all it was just too hazardous. NASA Administrator James Webb set up an accident review board on 3 February, which presented its 3000-page findings on 5 April. In summary, it reported: 'The fire in Apollo's floor was most probably brought about by some minor malfunction or failure of equipment or wire insulation. This failure, which most likely will never be positively identified, initiated a sequence of events that culminated in the conflagration.'

The investigations suggested that the fire probably began as a short circuit in the wiring near Grissom's couch, though no one could be sure of this. But undoubtedly a spark initiated by an electrical short-circuit was the source of the fire. Thereafter tragedy was inevitable. In the pressurized pure oxygen atmosphere virtually everything inside the cabin was potentially combustible – insulation, papers, clothing, plastics, and so on. And the fate of the astronauts was irrevocably sealed by the hatch design, which made a quick exit from the spacecraft impossible.

Kennedy Space Center director Kurt Debus, accepting the ultimate responsibility for the accident, said: 'It is for me very difficult to find out why we did not think deeply enough or were not inventive enough to identify this [the simulated countdown] as a very hazardous test.

'We never knew that the conflagration would go that fast through the spacecraft so that no rescue would essentially help. This was not known. This is the essential cause of the tragedy. Had we known, we would have prepared with as adequate support as humanly possible for egress.'

Getting back on course

The 5 April report delved deeply into the background to the accident and apportioned blame not only to the spacecraft manufacturers, North American Aviation (which later became Rockwell), for sloppy design and manufacture, but also to NASA for slipshod test procedures and less than adequate safety precautions. As a result, NASA's Apollo Spacecraft Program Manager Joe Shea was replaced by George Low. He chaired a group known as the Configuration Control Board, whose job it was to evaluate and approve changes in Apollo design to ensure that such a tragedy could never happen again.

Over the following months the Board considered nearly 1700 changes and ultimately approved 1341 of them. Many were

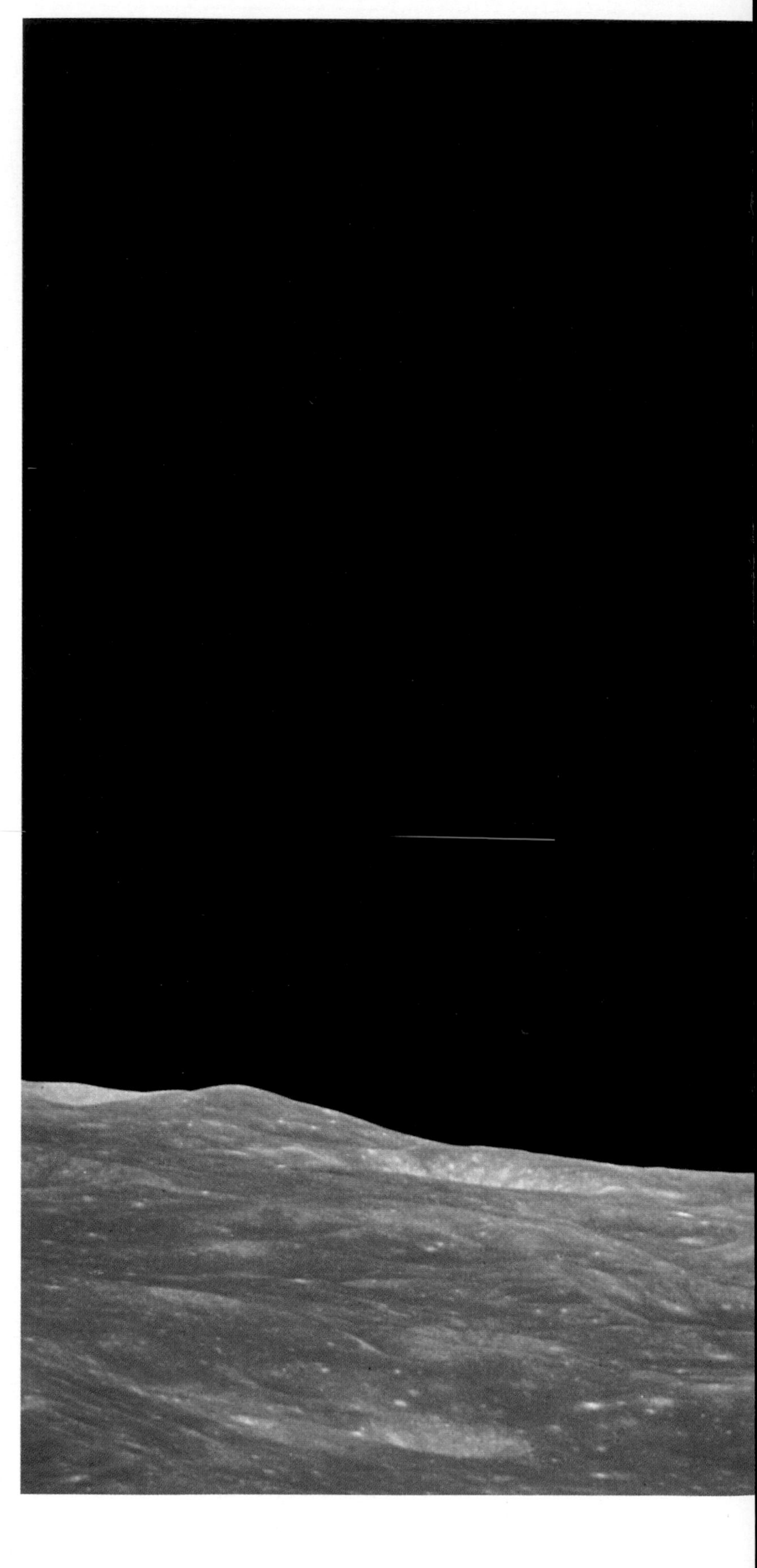

Apollo 8: Earthrise
During their 10 orbits of the Moon, the astronauts snap this famous picture of 'spaceship' Earth rising over the lunar horizon. The continent near the terminator – the boundary between light and shadow – is Africa. The South Pole is located in the white area to the left. Much of the Atlantic Ocean is clear, but North and South America are hidden by cloud.

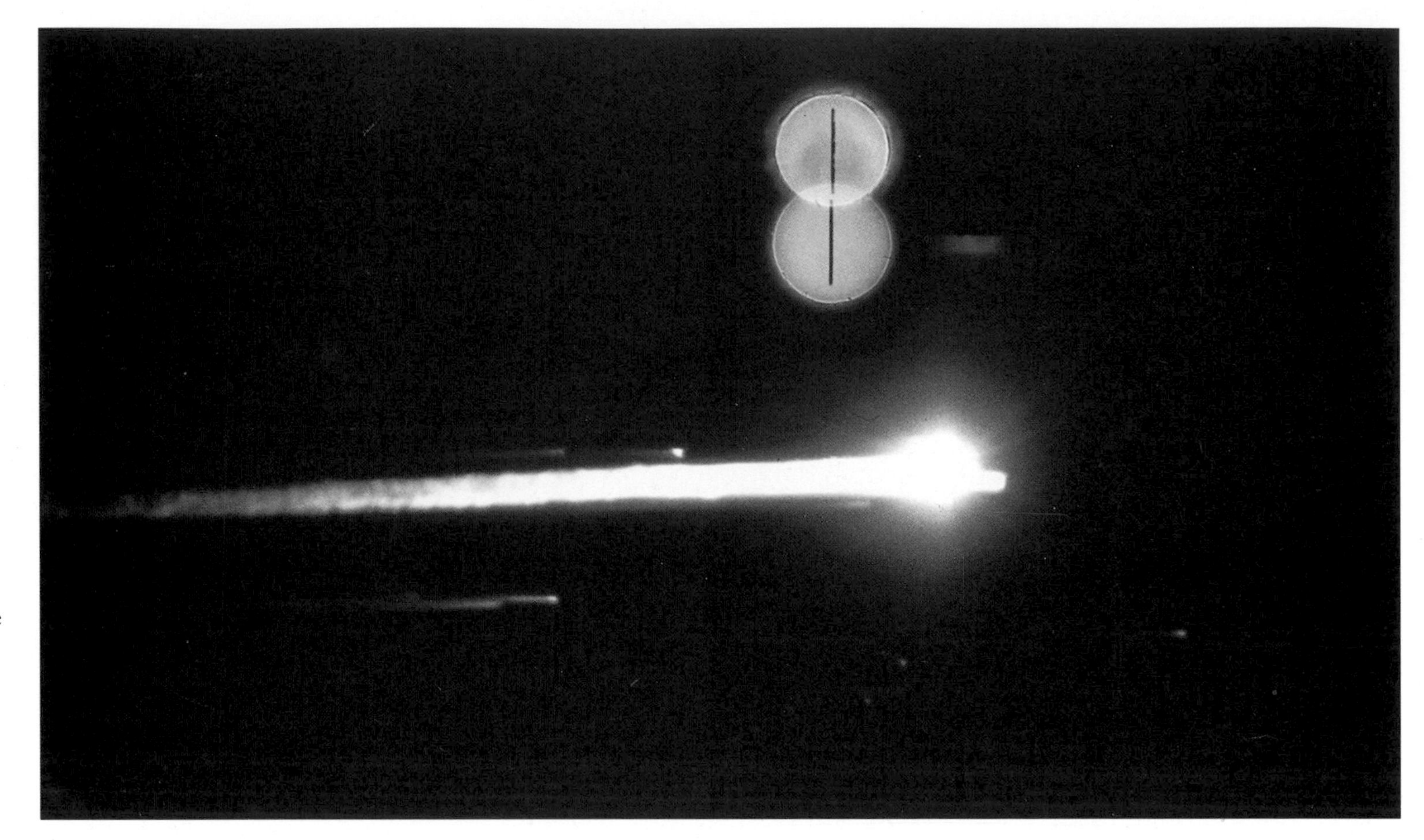

▶ **Apollo 8: Re-entry**
On 27 December 1968 comes potentially the deadliest part of the mission, re-entry into the Earth's atmosphere at a speed of nearly 40,000 km/h (25,000 mph). A special tracking camera mounted on a high-flying KC-135A aircraft captures this unique picture of the command module slamming into the upper air, heat-shield blazing.

fire-related. A new flameproof material called Beta Cloth replaced nylon in the spacesuits. Electrical wiring was rerouted and more adequately covered. The 16 psi cabin atmosphere while on the ground would be changed from 100 per cent oxygen to a 60/40 oxygen/nitrogen mix. After lift-off the pressure would be gradually reduced to 5 psi – one-third of an atmosphere – and switched to pure oxygen.

The hatch was redesigned completely to allow rapid crew egress. The former double-hatch bolt-secured system was replaced by a push-pull single outward-opening hatch. A crew member could unlatch the hatch in just three seconds; as could a white-room technician on the outside. From the white room astronauts and technicians could then exit rapidly down a slidewire, reaching the ground in seconds.

It would take fifteen months for the design changes to be implemented and for the first new Block II command module to be delivered to the Cape.

'Nothing will stop us'

While in early April 1967 the Apollo accident report was being analyzed in the United States, rumors came from the Soviet Union of an imminent new space spectacular. It would involve the rendezvous and docking of a new type of craft that would lead within perhaps 18 months to a manned flight to the Moon. On 22 April a new spacecraft was placed in orbit. The last craft designed by Korolev, it was named Soyuz, meaning 'Union'. On-board was a single cosmonaut, Vladimir Komarov. Subsequently, details emerged that this Soyuz 1 was intended to link up with Soyuz 2. The two ships would then perform maneuvers together, and there would also be a transfer of crews and spacewalking.

In the event this flight plan was scrubbed on 24 April for reasons unknown, and Komarov was recalled to Earth after only 18 orbits. He did not survive the always hazardous return to Earth. The official explanation for space's first in-flight casualty was that after a successful re-entry the lines of the main parachutes became entangled at an altitude of about 7 km (5 miles), making them ineffective. Unbraked, the descent module, with Komarov inside, smashed into the ground at high speed. He was killed instantly. Where Komarov crashed, at Orsk near the Urals, there is now a small shrine in his memory. Strangely, Orsk is far removed from the landing line for Soviet spacecraft. Although no official comment for this discrepancy has ever been offered, a major malfunction in one of the new Soyuz spacecraft's systems seems most likely.

It was now the turn of the Soviet Union to be stunned by tragedy. As Komarov's ashes were being interred in the Kremlin wall, pioneer cosmonaut Yuri Gagarin reflected: 'Mankind never gains anything gratuitously. There has never been a bloodless victory over nature.'

Later, summing up the resolve of his fellow cosmonauts, and undoubtedly also of their American counterparts, he wrote: 'Nothing will stop us. The road to the stars is steep and dangerous. But we're not afraid. Every one of us cosmonauts is ready to carry

◀ **Apollo 9: CSM *Gumdrop***
On 7 March 1969 *Gumdrop* coasts in Earth orbit, flying in formation with lunar module *Spider*, from which this picture is taken. On-board the CSM is pilot David Scott.

▼ **Apollo 9: LM *Spider***
David Scott in turn shoots this picture of the appropriately named *Spider*, piloted by James McDivitt and Russell Schweickart. *Spider* has its legs extended in the landing position, with sensors dangling beneath the foot pads. In a real landing these would signal touchdown.

on the work of Vladimir Komarov.'

Komarov's death threw a spanner in the works of the Soviet space machine just as surely as the Apollo 1 disaster had in the American. But like the US, the Soviets resolved to press on and to come first in the race to the Moon.

Back in business

In the US the Apollo fire had not halted the development of von Braun's Moon rocket, the Saturn V. By the end of August 1967 it had been rolled out to the pad. On 9 November the Kennedy Space Center reverberated to the thunderous roar of the gigantic rocket's massive engines for the first time. The unmanned flight, designated Apollo 4, took place spectacularly at dawn. It launched a dummy Apollo CSM into orbit, which then used on-board engines to boost it to an altitude of 40,000 km (25,000 miles). The command module was then powered back into the atmosphere to undergo a high-speed re-entry. It was recovered after a successful splashdown.

In January 1968 the lunar module was tested in orbit after a launch by a Saturn IB. In April the second Saturn V launch placed in orbit another complete Apollo spacecraft for extensive testing. Apollo program director Lieutenant-General Sam Phillips then recommended that the next Apollo flight, Apollo 7, should be manned.

By mid-August the first newly designed command module was on the pad at Complex 34 atop a Saturn IB. On the morning of 11

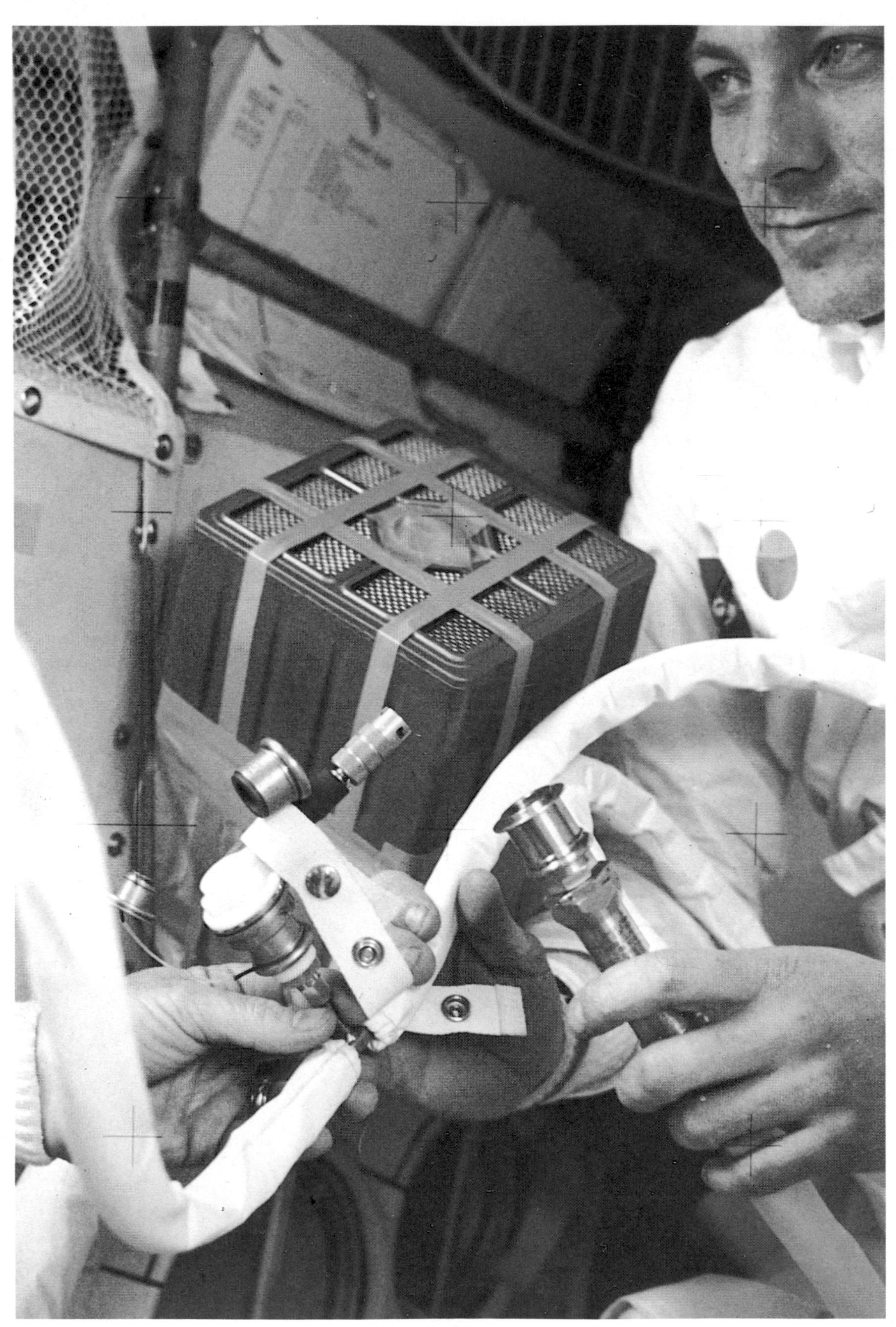

October, the Saturn's rockets fired to thrust Apollo 7 into orbit. Aboard were Walter Schirra, Walter Cunningham and Donn Eisele. From start to finish the first manned Apollo flight was all but flawless. The crew performed rendezvous and simulated docking with the spent third stage of their launch rocket. They tested ground-to-space communications links and made repeated firings of the service module's engine. The CSM performed like a champ.

The crew made several live telecasts during their 11-day, 163-orbit flight. And Earthbound viewers saw for the first time the strange effects of weightlessness. Despite developing head colds, the crew appeared in good spirits. They opened their first 'show' with a card that read: 'Hello from the lovely Apollo room high atop everything.' Later came another card: 'Keep those cards and letters coming in folks.'

However, behind the scenes, because of their colds and a punishing work-load, the astronauts became decidedly testy with Ground Control. Said commander Schirra at one point: 'I've had it up here today. We have a feeling that you down there believe some of these experiments are holier than God. We are a heck of a lot closer to Him right now.'

The splashdown of Apollo 7 – the 'Wally, Walt and Donn Show' as it was now dubbed – was eventful. As the command module hit the sea, the drag of the parachute lines flipped it upside-down. For nearly 20 worrying minutes the recovery ship some 14 km (9 miles) away lost contact with it. Then Schirra inflated air bags on the Apollo, and the craft righted itself.

Turtles round the Moon

As Apollo 7 came home, work was pressing ahead at the Cape on the next mission, which was to be not only the first manned Saturn V flight, but also a daring attempt to circumnavigate the Moon. It was scheduled for December 1968.

When this launch date had been announced the preceding August, it had taken the Soviet space team by surprise. They had been planning to get back in the Moon race by sending an unmanned spacecraft called Zond, a craft based on Soyuz, round the Moon first, while at the same time man-rating Soyuz again in Earth orbit. These plans had already been thwarted by the untimely death in a plane crash in March of Gagarin, who had been selected for the next Soyuz mission.

On 17 September they launched the unmanned Zond 5 spacecraft to the Moon. On-board were a live crew of turtles and fruit flies. Unlike earlier attempts with Zond craft, this one was successful. It looped around the Moon and sped back, re-entering the Earth's atmosphere on 21 September. It splashed down in the Indian Ocean, from which it was recovered. The turtles were still alive.

The flight had been closely monitored by Britain's giant radio telescope at Jodrell Bank. Jodrell's director Bernard Lovell startled the world by announcing that the telescope had picked up transmissions by a human voice! While assuming that it was a tape-recording for test purposes, he nevertheless reckoned that the next Zond flight would be manned.

On 25 October the unmanned Soyuz 2 craft sped into orbit, followed a day later by Georgi Beregovoi in Soyuz 3. It was just four days after Apollo 7 had splashed down. It seemed that the space rivals were truly running neck and neck in the Moon race. For five days Beregovoi carried out the maneuvers the ill-fated Komarov should have done, rendezvousing with Soyuz 2. He also gave a live telecast from orbit, and made a safe return to Earth. On 10 November, a week after Beregovoi's return, the unmanned Zond 6 headed away from Earth for another trip around the Moon. It came back on 17 November, landing on the ground this time close to the Baikonur launch site. A perfect mission. Preparations began immediately for Zond 7. This would be the one to carry man to the Moon. The world waited in anticipation. When would the Soviet Moon flight come? Would they beat the US, whose launch was scheduled for 21 December?

And God saw that it was good

The Soviet launch never materialized. So it was that on 21 December American astronauts set out to the Moon for the first time. On that day Frank Borman, James Lovell and William Anders became the first human beings to ride the Saturn V into the heavens; they became the first to break free of the Earth's gravity and journey to the Moon. On Christmas Eve they became the first to be captured by the Moon's gravity and enter lunar orbit.

During their 10 lunar orbits they made two live telecasts, showing to one of the biggest TV audiences ever what the Moon looks like close to. Lovell described the Moon as being 'essentially gray, no color. Looks like plaster of Paris. Sort of grayish sand.' Anders reckoned it was 'like dirty beach sand with lots of footprints on it'.

At the end of their final telecast during orbit nine of the Moon, Anders said: 'For all the people back on Earth the crew of Apollo 8 has a message that we would like to send you.' Then he, Lovell and Borman read out the hauntingly beautiful prose from the Book of Genesis, starting with 'In the beginning God created the Heaven and the Earth.' Borman ended with: 'And God saw that it was good,' concluding, 'And from the crew of Apollo 8 we close with goodnight, good luck, a Merry Christmas, and God bless you all – all

◀ **Apollo 13: Life-saving**
On the ill-fated unlucky 13th Apollo mission, the crew had to rig up makeshift apparatus to rid the air in the spacecraft of suffocating carbon dioxide gas. It was crude, but it worked. In the picture is John Swigert.

▼ **Apollo 13: Recovery**
The weary but, oh, so relieved crew emerge from the recovery helicopter on to the deck of the recovery ship USS *Iwo Jima* on 17 April 1970. From the left they are Fred Haise, James Lovell and John Swigert.

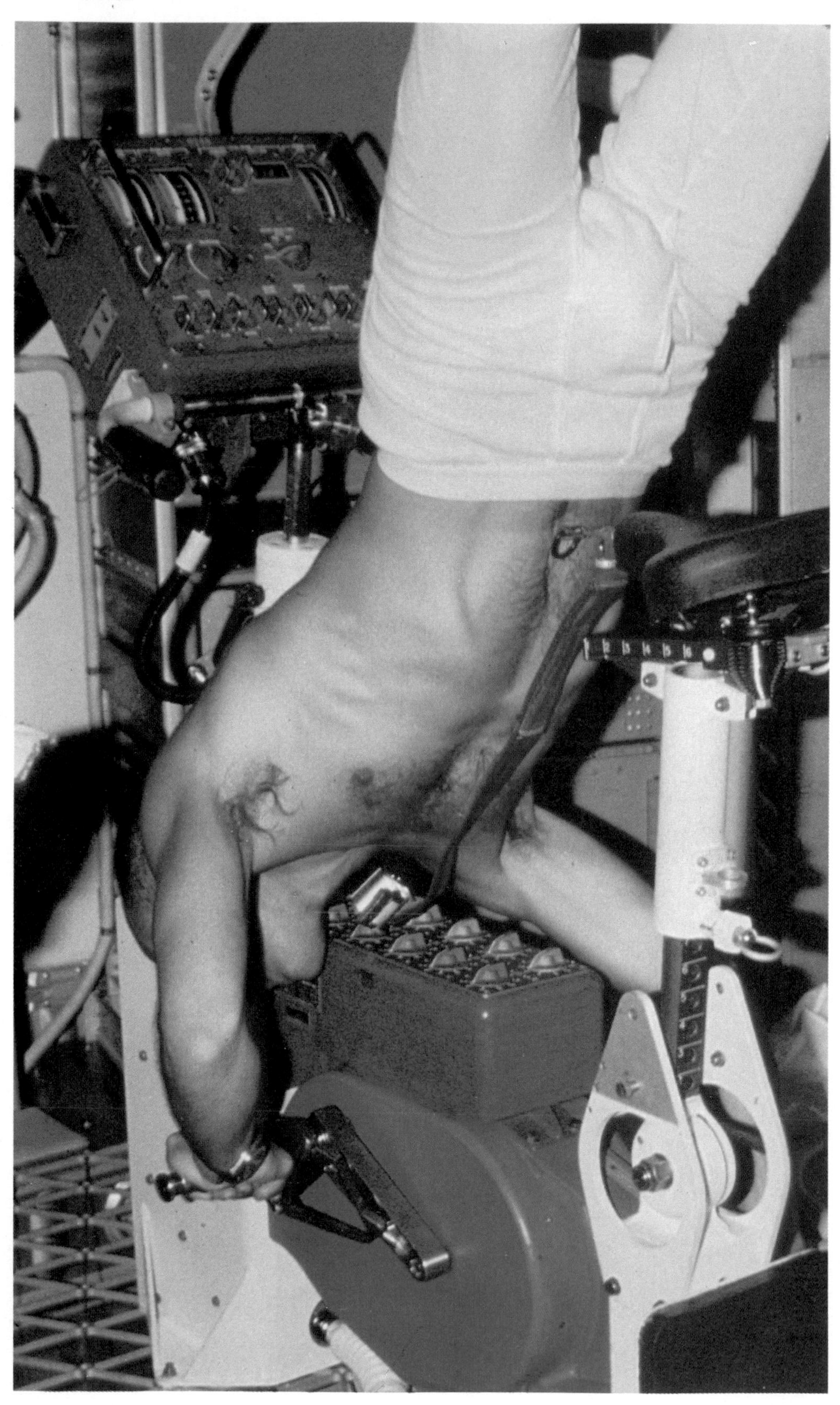

◀ **Skylab: Exercising**
Using redundant Apollo hardware, the US launches space station Skylab in 1973 to study among other things the long-term effects of weightlessness on the human body. Taking regular exercise in orbit proves vital. Here Skylab 2 astronaut Charles Conrad finds a novel way of pedaling the bicycle ergonometer.

of you on the good Earth.'

Apollo 8's return home was uneventful and as per textbook. The crew hit the re-entry window exactly right – if they had hit the atmosphere at the wrong angle, they would either have burnt up or bounced back into space, in either instance meeting their death. After their successful recovery, a helicopter pilot asked what the Moon was made of. Quipped Borman: 'It's not made of green cheese, it's made of American cheese!'

Certainly it wasn't Soviet cheese, for inexplicably the Soviet Moon effort fizzled out. No one really knows why. But it transpired that on 1 January 1969 the Soviet government declared that the main objective of the space program was to be the development of Earth-orbiting stations, supplied by Soyuz ferry craft. Later, Soviet spokesmen even went so far as to deny that they ever had any intention of launching a man to the Moon!

Gumdrop and *Spider*

With the decade fast running out, could NASA now bring off the impossible dream? It looked as if it could. With the year 1969 only three days old, the Saturn/Apollo 9 stack was trundled out to the pad, heading for a lift-off on the last day of February. A mild virus among the crew of James McDivitt, David Scott and Russell Schweickart put paid to that launch date, but a fit crew did take to the skies on 3 March. The Saturn/Apollo launch vehicle was in Moon-launching configuration for the first time – CSM on top, surmounted by an escape rocket, and lunar module underneath inside a shroud in the third rocket stage. The prime objective on the mission was to put the lunar module through its paces and go through the maneuvers necessary to configure the Apollo spacecraft for a lunar landing role.

After reaching orbit, Scott separated the CSM from the third stage of the launch vehicle and, firing the RCS thrusters, turned round and then headed back to dock with the lunar module still inside the third stage. He then edged the now complete Apollo spacecraft away from the third stage. On the third day of the mission McDivitt and Schweickart crawled into the lunar module and checked out its systems. A planned spacewalk by Schweickart was postponed because he was suffering from space sickness. Next day, however, he was well enough to stand on the porch of the lunar module, with his feet secured on restraints prosaically termed 'golden slippers'. Exclaimed Schweickart: 'Boy, oh Boy. What a view!' The main purpose of this stand-up EVA was to check the spacesuit and backpack, or PLSS (portable life-support system) which astronauts would use on the Moon.

On the fifth day of the mission McDivitt and Schweickart powered up the lunar module and undocked from the CSM. The

▲ (left) **ASTP: Docking**
This artist's impression shows Apollo and Soyuz spacecraft docked together in orbit on 17 July 1975. They remain linked up for 44 hours. For Apollo it is the swan song, but Soyuz is still shuttling into orbit 14 years later.

▲ **ASTP: Training**
The availability of a redundant Apollo CSM allows the US to plan a joint mission with a Soviet Soyuz spacecraft. Here astronaut Vance Brand (left) and cosmonaut Valery Kubasov are pictured in training for the flight at the Johnson Space Center at Houston.

modules now had their own callsigns, the CSM *Gumdrop*, the lunar module *Spider*. McDivitt fired *Spider*'s descent engine to separate from *Gumdrop,* and the two craft drifted some 80 km (50 miles) apart. After a second descent engine burn, the descent stage of the two-section craft was cut loose, the astronauts being in the ascent stage. They drifted to a maximum separation of some 185 km (115 miles) from *Gumdrop*, at which time they were out of visual contact. *Spider*'s on-board computer, radar and other instruments then worked out and effected the necessary ascent engine burns to rendezvous with *Gumdrop.*

It all worked perfectly. After some six hours separation Scott in *Gumdrop* spied *Spider* coming into dock and pronounced it 'the biggest, friendliest funniest looking spider I've ever seen'.

Snoopy and *Charlie Brown*

Apollo 9 splashed down after its flawless 10-day mission on 13 March. Two days earlier the hardware for the next and final Apollo test flight had set off at a snail's pace for launch pad 39B at the Cape. It was the first time that this pad, a twin of 39A, had been used. This upcoming mission, Apollo 10, was scheduled for May. The astronauts were to fly to the Moon and carry out a simulated lunar landing, but descending only to an altitude of some 15 km (9 miles). It was to be a full dress rehearsal for the first attempt at a lunar landing by Apollo 11 in July.

The crew, Thomas Stafford, Eugene Cernan and John Young, had what they termed 'a rocky ride' on the Saturn V into Earth orbit on 18 May. They entered lunar orbit three days later. On their 12th orbit of the Moon Stafford and Cernan transferred to the lunar module *Snoopy* and separated from Young in the CSM *Charlie Brown*.

After flying in formation for some time conducting systems checks, *Snoopy* swooped down to a tantalizing hair's-breadth of the lunar surface. 'Ah Charlie,' Cernan sighed, 'we just saw Earthrise and it's got to be magnificent.' Commented Stafford: 'There's enough boulders around here to fill up Galveston Bay too.' As they flew along the approach path (which they called the US1) that Apollo 11 would follow two months later, they described recognizable features.

Soon the time came for them to jettison the descent stage of the lunar module and fire the ascent-stage engine to return to rendezvous with *Charlie Brown.* As they effected the jettison maneuver, known as staging, *Snoopy* began to gyrate wildly. 'Son of a bitch!' exploded Cernan, fighting to regain control, which he did after about three minutes. 'I don't know what the hell that was, babe,' he later reported to Mission Control. In fact it appeared that a switch had inadvertently been left in the wrong position. The moment of panic passed. And the ascent-stage burn, and rendezvous and docking maneuvers were carried out without incident.

The crew arrived back home in triumph on 26 May after an eight-day mission, which suggested that, barring the unforeseen, a Moon-landing was on the cards come July. As Apollo 10 splashed down, a sign went up in the Mission Control room at Houston. It read '51 days to go'. When next a Saturn V blasted into the sky, it would be carrying Earthlings in the Apollo 11 spacecraft on the most incredible journey in history, to plant human footprints on an alien world for the very first time.

The one that didn't make it

The story of the Moon landings is recounted in Chapter 6, 'Walking on the Moon'. It was Neil Armstrong who had the honor of

▼ ASTP Soyuz: Lift-off
Soyuz blasts off the launch pad at Baikonur Cosmodrome on 15 July 1975, heading for a historic rendezvous in orbit with an Apollo spacecraft. The launch vehicle, known as the SL4 or A2, has been the standard launcher for Soyuz since 1967. It will continue to be through the 1980s.

▶ ASTP Apollo: In orbit
An excellent picture of the American spacecraft taken by Soviet cosmonauts in Soyuz. It shows well the docking module attached to the Apollo command module. At the free end of the docking module is a port that will receive the Soyuz docking probe.

becoming the first man-on-the Moon, stepping down on to the surface on 20 July 1969. The American people had achieved the seemingly impossible task set by President Kennedy. They had landed a man on the Moon and returned him safely to Earth with five months of the decade still to run. For good measure they repeated the feat with Apollo 12 in November, again before the decade was out.

With another mission scheduled for 11 April 1970, it seemed as if traveling to the Moon was becoming routine. That mission, Apollo 13, would be aiming for lunar landfall at Fra Mauro in a rugged region of the Ocean of Storms. Just as the final countdown began, Charles Conrad, one of the back-up crew, went down with rubella (German measles). Tests were immediately carried out on the prime crew, James Lovell, Fred Haise and Ken Mattingly. Of these, it was found that Mattingly had no resistance to the disease. With only one day to go before launch, NASA decided that it would be foolish to risk him developing the disease during the flight. So they pulled him out and put in John Swigert instead, who possessed the necessary immunity.

Apollo 13 blasted off on time on 11 April. Said Kennedy Launch Control as they set off: 'Good luck, and head for the hills!' But luck for this crew soon ran out. For 55 hours the mission went with characteristic smoothness. Lovell was winding up a telecast from space on which he had given viewers a conducted tour of their spaceship. 'This is the crew of Apollo 13 wishing everyone there a nice evening,' he said. 'We're just about ready to close out our inspection of *Aquarius* [the lunar module] and get back for a pleasant evening in *Odyssey* [the CSM].'

But a pleasant evening it was certainly not going to be. Suddenly,

▼ **ASTP Soyuz: Rendezvous**
The Apollo crew photograph the Soviet spacecraft as the two craft prepare to link up with one another on 17 July 1975. They will remain docked for nearly two days.

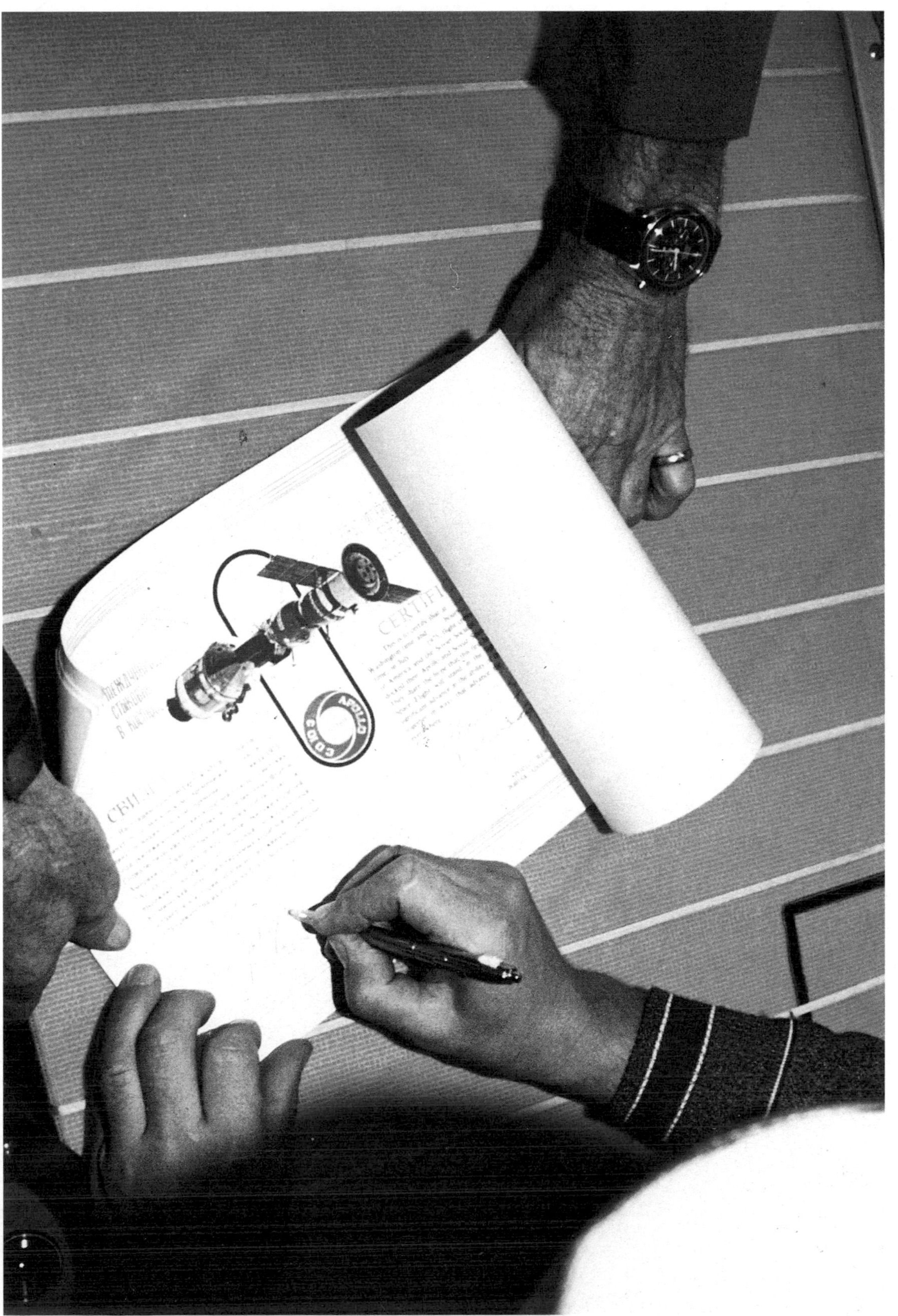

▶ **ASTP: Signing ceremony**
Inside the Soyuz orbital module, Valery Kubasov adds his signature to the official certificate drawn up to mark the historic orbital rendezvous of two great spacefaring nations.

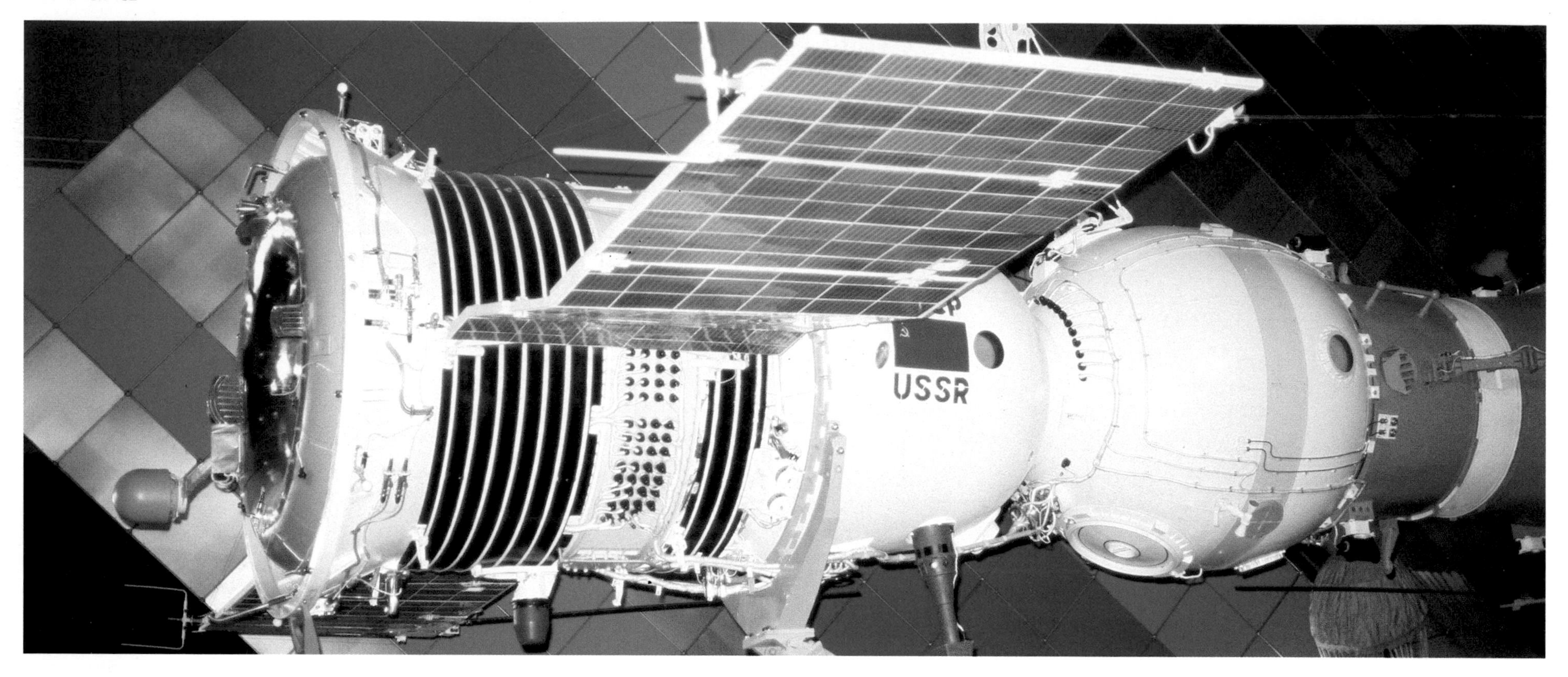

the crew heard 'a pretty large bang' in *Odyssey*. Scanning the instruments Swigert saw the electrical supply falling rapidly. 'Okay, Houston,' he reported to Mission Control, 'we've had a problem here.' It was a masterly understatement.

An explosion in the liquid oxygen tank in the service module had knocked out the power system and oxygen supply to *Odyssey*. The CSM was effectively dead. It had a battery pack, but that had to be reserved for re-entry maneuvers. It had only a 10-hour life anyway, and the crew were, even at the most optimistic calculations, at least 90 hours from home. There was only one way the crew could survive, and that was to use the intact *Aquarius* as a lifeboat, using its power and oxygen supply for life support and its descent-stage engine to alter their trajectory so that they looped once around the Moon and returned to Earth.

In the 'lifeboat mode' *Aquarius* did what was demanded of it and, against the odds, brought home the valiant crew – cold, weary, but alive. On 17 April, as Apollo 13 accelerated towards the Earth's atmosphere, the crew transferred to *Odyssey* and cut *Aquarius* loose. Radioed Mission Control: 'Farewell *Aquarius* and we thank you.' *Odyssey* survived re-entry and splashed down safely. NASA afterwards described the mission as 'the most successful failure in the annals of space flight'. Commander Lovell, speaking later to members of the Kennedy launch team, remarked: 'I think the mission matured the space program a little, because people were perhaps getting a bit complacent about what we do.'

Spin-off from Apollo

It was January 1971 before the next lunar landing was attempted, on mission Apollo 14. That aimed for, and reached, Fra Mauro. Three more successful landing missions took place, ending with Apollo 17 in December 1972. There were originally to be other missions, but these were cancelled, leaving a stock of redundant hardware. Some of this was soon put to good use in the Skylab project of 1973/4 and the Apollo-Soyuz Test Project of 1975.

Skylab was a project that used Apollo derived hardware to build a large experimental space station, and a surplus Saturn V vehicle to launch it. It was not the first space station in orbit, however. The Soviets had launched this in April 1971. This space station, Salyut 1, was not a great success and was attended in June 1971 by another in-flight tragedy, when three cosmonauts returning from it in Soyuz 11 were killed. Coming just over a year after Apollo 13's near disaster, it underlined again just how dangerous space flight really is. For an account of the Skylab and Salyut space-station missions, see Chapter 7.

Cooperation in the cosmos

The Apollo-Soyuz Test Project had its origins in meetings between representatives of NASA and the Soviet Academy of Sciences, which began in Moscow in October 1970. The attendees agreed on a joint space venture to take the form of a rendezvous and docking mission in orbit. President Nixon and, for the Soviets, Alexei Kosygin, in May 1972 approved the idea of a joint Apollo-Soyuz flight as part of an agreement concerning 'Cooperation in the exploration and use of space for peaceful purposes.'

The crews selected for the mission were, for the US: Thomas Stafford, Vance Brand and Donald Slayton. For Slayton it was effectively the last chance of getting into space. One of the

▲ **The Soyuz spacecraft**
The three sections of this three-module spacecraft can be clearly seen in this exhibition model. To the right is the spherical orbital module and to the left is the instrument module with its attached 'wings' of solar panels. In between is the descent module, in which the cosmonauts return to Earth.

▶ **Soyuz: Off to launch**
An unusually exotic multi-exposure picture of the upper end of the SL4/Soyuz launch vehicle. It shows the vehicle starting to be upended into its launch position. At the top of the vehicle is the emergency escape system, which would pull the spacecraft clear from the rocket in a launch emergency.

'Original Seven' astronauts, he had been grounded because of a heart condition through Gemini and Apollo. The Soviet crew comprised Alexei Leonov, who had made the world's first spacewalk in 1965, and Valery Kubasov.

The crews trained at each other's facilities at the Yuri Gagarin Cosmonaut Training Center in Star City near Moscow, and at the Johnson Space Center in Houston. So that the two craft could link up in space, a cylindrical docking module was designed with different docking mechanisms at each end – one for Apollo, one for Soyuz. The module would also act as a kind of acclimatization airlock to act as a buffer between the low-pressure pure oxygen atmosphere of Apollo and the atmospheric pressure nitrogen/oxygen atmosphere of Soyuz. Obviously, the language barrier was going to be a problem so both crews learned a smattering of the other's language. During the joint phase of the flight the American crew were to speak Russian, the Soviet crew, English. A bilingual instruction book was to be carried by both spacecraft.

The mission was scheduled for 15 July 1975, and both spacecraft lifted off on time. Soyuz blasted into orbit first from Baikonur Cosmodrome, followed some seven hours later by Apollo from the Kennedy Space Center. A Saturn IB launched the Apollo. As the spacecraft went into orbit, Slayton, at 51 the oldest man so far to fly in space, said: 'Man I tell you, this is worth waiting 16 years for!' Once in orbit Apollo had to turn around and dock with the 3-meter (10-foot) long docking module.

On 17 July the two craft maneuvered close to one another. Said Stafford in Russian: 'Soyuz is very beautiful.' At 12.09 Eastern Daylight Time (EDT) the two craft linked up above the Atlantic Ocean west of Portugal. Said Stafford (in Russian): '... we have succeeded. Everything is excellent.' Commented Leonov (in English): 'Soyuz and Apollo are shaking hands now,' and later, 'Well done Tom, it was a good show.' Soon the two crews met inside the docking module and shook hands. It was more than just a symbolic handshake because all had become close friends.

For 44 hours the two craft remained docked. The crews ate together and exchanged mementos; they also carried out a number of scientific experiments. Two days after separation Soyuz returned to Earth. For the very first time, its landing was televised live, as its launch had been. The return of Apollo on 24 July was dramatic. For a start the command module flipped upside-down on hitting the water. Then yellow gas began seeping in. It was deadly nitrogen tetroxide, a gas used in the reaction control thruster system. Stafford eventually wriggled free of his couch straps and reached the stowed oxygen masks. But by the time he got to Brand, Brand was unconscious through inhalation of the fumes. He regained consciousness after a minute or so. After recovery, the crew were checked out but found not to have suffered any permanent lung damage.

So with this potentially fatal return to Earth, the Apollo program and the spin-offs from it came to an end. It was for the US the end of the expendable era of manned spacecraft. Next time American astronauts ventured into orbit – and this would not be for nearly six years – they would do so in the world's first reusable space vehicle, the space shuttle. However, Soviet cosmonauts would continue to ride into orbit in Soyuz through the 1980s. Their first reusable space plane *Buran* did not make its maiden and unmanned flight until the fall of 1988.

◀ Soyuz: On the pad
The spacecraft has just been carried horizontally to the pad by a rail transporter and erected into the vertical position. Soon the arms of the jawlike gantry will swing up to encircle the rocket and provide access for the technicians who will complete the launch preparations.

▲ Soyuz: Landing
The cosmonauts re-enter the Earth's atmosphere inside the descent module. Braking by atmospheric drag, retrorockets and parachutes slow them down to a gentle touchdown, always on land. Afterwards, while waiting for transport back to base, the cosmonauts as a rule autograph the descent module.

Chapter 3

SHUTTLING INTO ORBIT

◀ **STS-1: *Columbia***
The fully assembled shuttle stack begins its long journey to the launch pad as the doors of the Vehicle Assembly Building open. The date is 29 December 1980.

▶ **STS 51-J: *Atlantis***
The maiden launch of the fourth operation orbiter, *Atlantis*, on 30 October 1985. Five rockets combine to thrust the incomparable flying machine into the blue.

On Sunday, 12 April 1981, a few seconds after 7 am on a perfect Florida morning, the reclaimed swampland that is the Kennedy Space Center reverberated with a deafening roar as a new breed of space vehicle took to the air for the first time. It was a gleaming white and black rocket plane riding piggy-back on a large tank, with twin boosters alongside belching thick columns of flame and smoke. The rocket plane was named *Columbia*; the event was the maiden flight of the space shuttle. *Columbia*'s was a perfect maiden flight. And only seven months later it was punching its way back into the heavens. No space vehicle had ever done this before.

Columbia was the first of a fleet of orbiting space planes, or orbiters, able to return repeatedly to space. The orbiters are the key components in the reusable space shuttle launching system. The shuttle system has other advantages besides reusability. It can lift huge payloads into orbit, which can be almost as big as a railroad car and as heavy as four African elephants. It is user-friendly, allowing ordinary healthy people – not just highly trained astronauts – to journey into space. This is because its crew cabin is pressurized like an airliner's and its acceleration forces on lift-off (3gs) are comparable with those on a terrestrial roller-coaster ride.

In this chapter we see how the shuttle operates and trace its birth, and rebirth in its 26th flight after being grounded for 32 months. By then it had acquired a rival, a Soviet lookalike called *Snowstorm*.

United States
USA

PRESIDENT RICHARD NIXON gave the go-ahead for the space-shuttle program early in 1972, when the greatest adventure of all time – the Apollo exploration of the Moon – was drawing to a close. He gave his approval in a speech he made on 5 January.

'The United States', he said, 'should proceed at once with the development of an entirely new type of space transportation system designed to help transform the space frontier of the 1970s into familiar territory, easily accessible for human endeavor in the 1980s and '90s.... It will revolutionize transportation in near space by routinizing it.... It will take the astronomical costs out of astronautics.... This is why commitment to the space shuttle's program is the right next step for America to take.'

By that time the essential features of the present shuttle system – reusable orbiter, recoverable solid rocket boosters and expendable external tank – had been decided. But many other options had been considered in the preceding years and dropped because of the tight development budget.

Rocket planes

There was no shortage of designs for a reusable space launch system. In the 1940s and 1950s respected authorities such as Wernher von Braun, architect of the Saturn V Moon rocket, pointed out the benefits of reusable craft as Earth-to-orbit cargo carriers. Following America's successful thrust into orbit in January 1958, the American Rocket Society presented an influential paper entitled: 'Commercial rocket airplane: A connecting link to manned space flight'.

And in effect a rocket airplane was – the legendary X-15. This slim dart of a plane was dropped from a B-52 carrier aircraft at a

▸ ***Columbia:* Assembly**
Sticking the 34,000 tiles in position proves a daunting task. Each one is individual, identified by a computer-coded number. Development problems with the tiles are a major factor in delaying the first flight attempt of the space shuttle.

▾ ***Enterprise:* Test flight**
Flight tests in the atmosphere of the prototype orbiter *Enterprise* begin on 12 August 1977. Here Fred Haise and Gordon Fullerton fly the orbiter for the first time over the Mojave Desert in California, separating from their 747 carrier aircraft and gliding to a runway landing.

height of about 14,000 meters (45,000 feet). It then fired its rocket engines to soar to the fringes of the atmosphere and then glide into a high-speed, unpowered landing. The X-15 flights, centered upon the Edwards Air Force Base in California, began in 1959. They took eight of their pilots above an altitude of 80 km (50 miles), so high that they were given astronauts wings.

There were plans to develop an orbital version of the X-15, but they were abandoned in favor of the cheaper expendable 'capsule on a missile' spacecraft, typified by Mercury (see page 38). The 99 X-15 flights, however, yielded voluminous data on high-speed flight at up to nearly seven times the speed of sound (6800 km/h, 4250 mph) and at altitudes up to 107 km (67 miles).

Another advanced American manned space flight concept involved the use of a kind of mini-shuttle, which would be launched into space by an expendable rocket and return in gliding flight. But this project, called Dynasoar ('dynamic soaring'), was also cancelled.

Reusability

NASA, in the rush to get to the Moon within the decade of the 1960s, had little time to develop the concept of the space plane. Then Neil Armstrong set foot on the Moon, and it was time to start thinking about future needs. In September 1969 it recommended to the President that immediate priority be given to the

◀ **On the second attempt,** *Columbia* blasts off the launch pad, climbing out of the clouds of flame, smoke and steam that envelop the launch pad.

▼ **Jubilation in Kennedy** Launch Control Center after *Columbia's* successful launch.

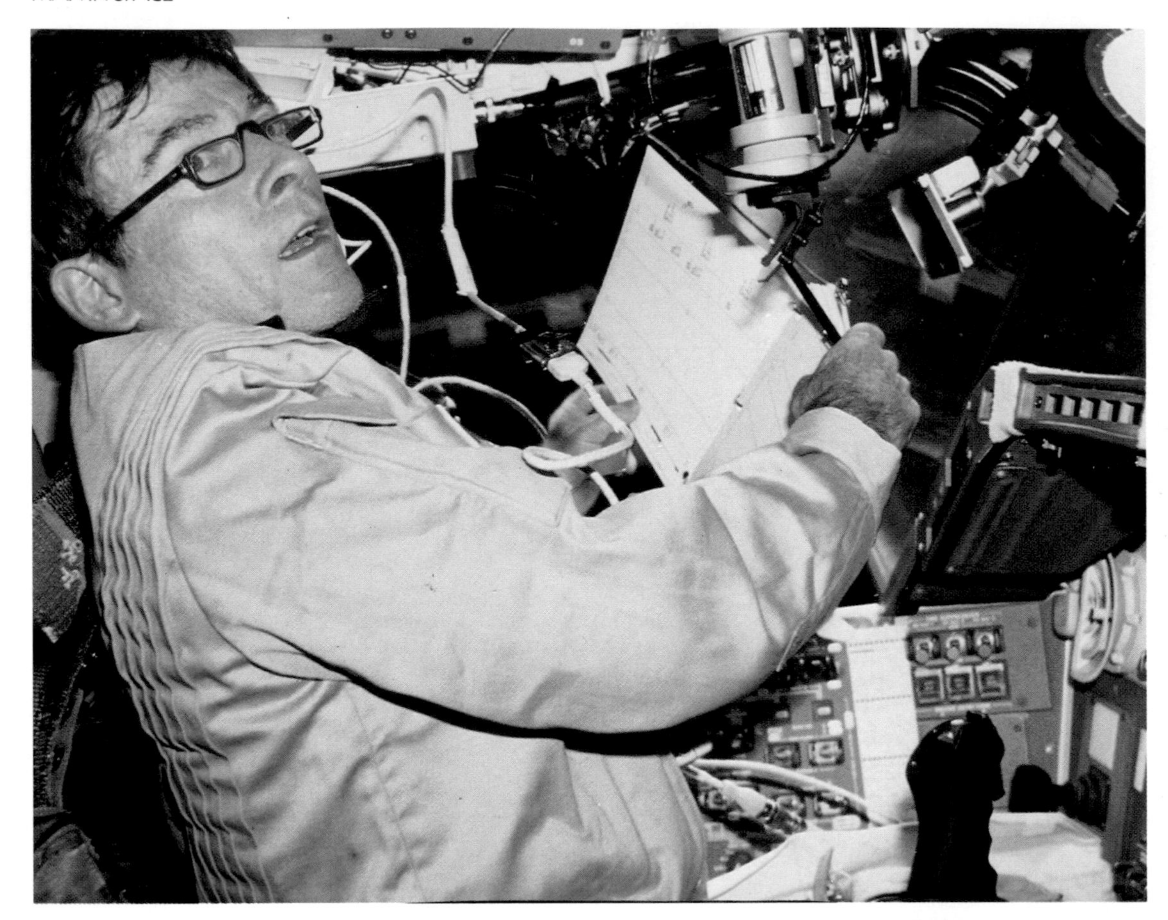

development of a reusable launch vehicle as the next major step in the space program after Apollo.

In March 1970 President Nixon gave his approval, and serious design studies began. At first the designs focused on a fully reusable two-stage system. The first stage would take off vertically and lift the second stage to high altitude. It would then return to land on a runway. The second stage meanwhile would fire its own engines to boost it into orbit.

However, the design would be expensive to develop, and the idea of a manned first stage was abandoned in favor of unmanned but recoverable boosters. The design of the orbiter also underwent major changes. These resulted in a small delta-winged plane, which had its own engines but drew fuel from an external tank. One of the final decisions in the basic shuttle design centered on the boosters – would they be liquid- or solid-propellant motors? Liquid-propellant boosters were more powerful, but were more expensive and would be more difficult to recover undamaged. Solid boosters on the other hand would be cheaper and easier to recover, but had never before been used for manned space flight.

Good *Enterprise*

The solid rocket boosters won out and on 12 April 1981 were used to carry men into space for the very first time. The occasion was the first flight of the space shuttle, which officially is termed the Space Transportation System (STS).

The first mission was designated STS-1. It featured the first operational orbiter, *Columbia*. On that mission veteran astronaut John Young and rookie Robert Crippen flew very much as test pilots. The shuttle was designed as a manned system. There was no way of flight testing it other than to do it for real. Each individual

▲ **STS-1: In-orbit activity**
Up in space John Young catches up on some paperwork during the spectacularly successful 54½-hour mission. He is pictured in the commander's seat on the left of the cockpit.

▶ **STS-1: Landing**
Moments before touchdown as the orbiter comes in to land at Edwards Air Force Base in the Mojave Desert on 14 April 1981. It is a perfect landing. Comments Robert Crippen: 'What a way to come to California!'

part of the system had been tested by itself as far as was possible, but never had the whole integrated assembly been flown.

That maiden flight had been a long time coming. It was at least two years late. Construction of a prototype orbiter, called *Enterprise* after *Star Trek*'s illustrious spaceship, was completed in 1977, and then underwent flight tests in the atmosphere to verify the orbiter's gliding performance. In the first test (February 1977) *Enterprise* was mounted on a converted Boeing 747 jet and flown in that configuration. On later tests, starting in August 1977, *Enterprise* separated from the carrier jet to glide down to a runway landing at Edwards Air Force Base in California.

By the completion of the approach and landing tests in October 1977, the aerodynamic characteristics of the orbiter design had been proved excellent. And the first space flight was scheduled for some time in 1979. But it was not to be, because of development problems with the all-new main engines and the ceramic tiles that formed the heat shield. The tiles – there were over 33,000 of them on *Columbia* – were fragile and difficult to stick on. The maiden flight was thus delayed until April 1981.

Hail *Columbia*!

The first attempt to launch *Columbia* that April took place on the 10th. The excitement at the Kennedy Space Center, and indeed all along the Space Coast, was intense. No American had been launched into orbit since the Apollo-Soyuz Test Project six years before. Crowds gathered in their hundreds of thousands at every

▲ **STS-1: Mission Control**
Their own duties all but over, the staff at Mission Control, Houston, watch the TV monitors as *Columbia* makes a triumphant return to Earth.

◀ **STS-2**
As *Columbia* is readied for a second journey into space in November 1981, roadside messages along the Space Coast urge it on.

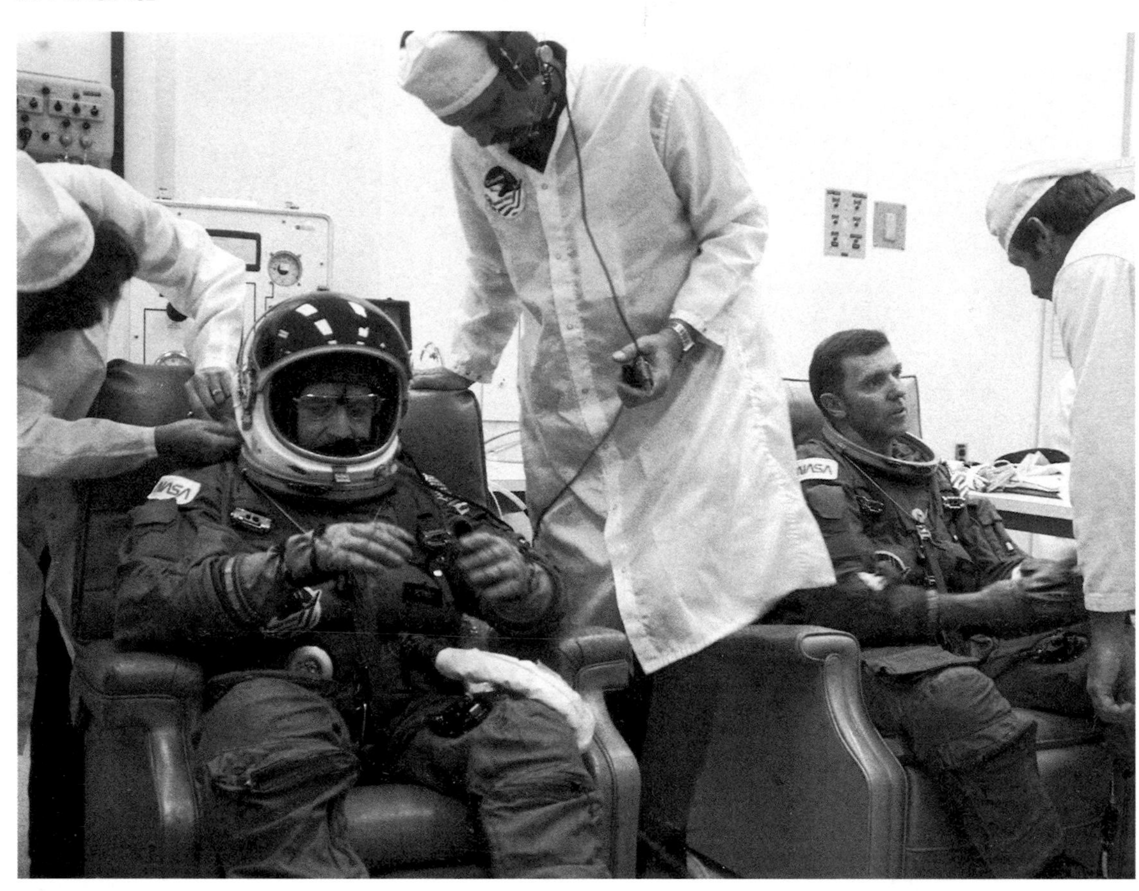

vantage point at the Center and in the surrounding areas.

The excitement rose to fever pitch as the countdown proceeded smoothly to a 6.50 am lift-off. Then the countdown clock stood at T-9 minutes. At this time there was a built-in hold (see page 86) for 10 minutes. Inside *Columbia*, Young and Crippen began bracing themselves for the fury that would be let loose when the rockets fired beneath them. Outside, the tension was palpable. But inexplicably the countdown clock stayed at T-9. The on-board computers that are the brains behind every shuttle flight malfunctioned. The technical name for the malfunction was 'a timing skew', which in popular parlance meant that the computers weren't 'talking' to each other. After vain attempts at reprogramming, the launch attempt was scrubbed.

Later that day NASA announced that the problem had been resolved and scheduled a new launch attempt for 7 am Sunday morning, 12 April. This time there were no glitches. The computers continued to talk to each other into the final countdown. Then they triggered off the main engines and solid rocket boosters. And *Columbia* lifted off, on time, to the accompaniment of the wildest and most emotional scenes of excitement witnessed at the Cape since Apollo 11 set off on the first Moon-landing mission.

Happy anniversary

Columbia's launch that April day was a triumphant demonstration of man's determination to explore the unknown, to go where no man had gone before. It was a fitting celebration of another epoch-making event that had occurred on that date 20 years

▲ **STS-2: Suiting-up**
In Kennedy's Operations and Checkout Building, Richard Truly (left) and Joe Engle are helped into their flight suits for the shuttle's second mission.

▶ **STS-2: Lift-off**
On 14 November 1981 history is made as a spacecraft returns to space for a second time. *Columbia*'s second flight is a two-day, 36-orbit repeat of the first.

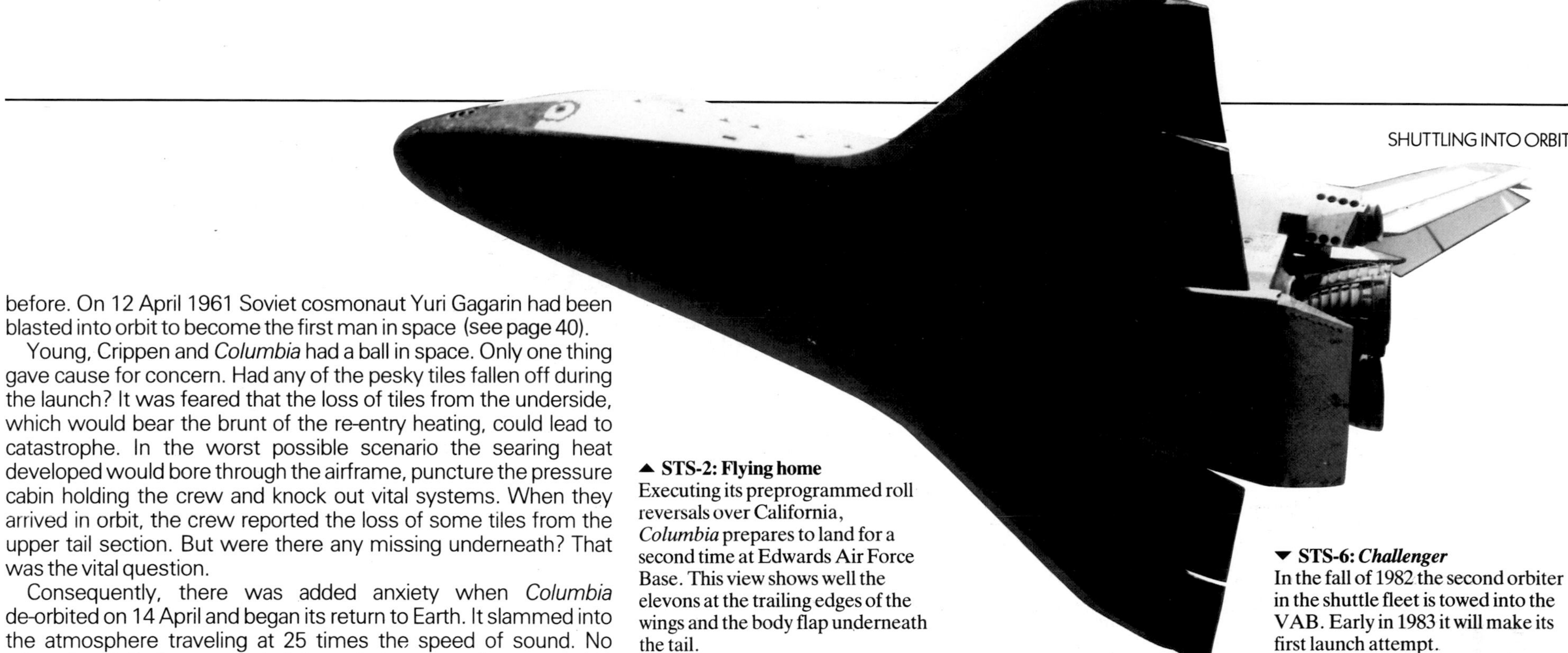

before. On 12 April 1961 Soviet cosmonaut Yuri Gagarin had been blasted into orbit to become the first man in space (see page 40).

Young, Crippen and *Columbia* had a ball in space. Only one thing gave cause for concern. Had any of the pesky tiles fallen off during the launch? It was feared that the loss of tiles from the underside, which would bear the brunt of the re-entry heating, could lead to catastrophe. In the worst possible scenario the searing heat developed would bore through the airframe, puncture the pressure cabin holding the crew and knock out vital systems. When they arrived in orbit, the crew reported the loss of some tiles from the upper tail section. But were there any missing underneath? That was the vital question.

Consequently, there was added anxiety when *Columbia* de-orbited on 14 April and began its return to Earth. It slammed into the atmosphere traveling at 25 times the speed of sound. No winged craft had ever traveled through the atmosphere at such a speed. Because of the usual communications blackout that occurs during any re-entry, no one but the crew knew for 21 minutes whether *Columbia* had survived re-entry intact. It had. Crippen observed later that during re-entry '*Columbia* was flying smoother than an airliner. Not a ripple!'

Soon *Columbia* was gliding in steeply to land at Edwards Air Force Base. It was a textbook landing. Commented Crippen afterwards: 'If you can imagine the smoothest landing you've ever had in an airliner, ours was at least that good. John really greased it in.' Said Mission Control when they touched down: 'Welcome home, *Columbia*. Beautiful, beautiful.'

It had indeed been a beautiful flight, which vindicated the space shuttle concept. *Columbia*, which had on occasions been disparagingly referred to as the Flying Brickyard, the Taj Mahal and the swept-back penguin, proved to be a 'magnificent flying machine'.

By November *Columbia* was back on the launch pad aiming for an unprecedented second trip into space. Riding the orbiter on this mission (STS-2) were Joe Engle and Richard Truly. On the designated launch day, 4 November, the countdown went smoothly down to T-9 minutes. Then the computers began to detect anomalies in some systems. Eventually, the countdown clock was restarted and ticked right down to 31 seconds before lift-off. Then the computers again called a halt to the proceedings, and the launch was scrubbed. Not until 10 days later did *Columbia* return to space.

Over the next year *Columbia* repeated this feat another three times. On its fifth flight (STS-5), on 11 November 1982, it became officially operational and carried into orbit two communications satellites. It also carried a crew of four, the most ever carried into space at one time.

After STS-5 *Columbia* enjoyed a well-earned rest. When it next blasted into orbit, it would be carrying a novel cargo, a fully equipped space laboratory called Spacelab (see page 194). Meanwhile its sister craft *Challenger* was to carry the American flag into orbit. It made its space debut on 4 April 1983. In the

▲ **STS-2: Flying home**
Executing its preprogrammed roll reversals over California, *Columbia* prepares to land for a second time at Edwards Air Force Base. This view shows well the elevons at the trailing edges of the wings and the body flap underneath the tail.

▼ **STS-6: *Challenger***
In the fall of 1982 the second orbiter in the shuttle fleet is towed into the VAB. Early in 1983 it will make its first launch attempt.

following year the third orbiter, *Discovery*, became operational (30 August 1984), followed in 1985 by the fourth, *Atlantis* (30 October 1985). But the shuttle fleet numbered four for only a few weeks longer. On 28 January 1986 *Challenger* exploded during launch (see page 132). President Reagan subsequently approved the construction of a replacement orbiter, which should become operational in 1991.

The versatile orbiter

The crew-carrying part of the shuttle, the orbiter, is the most advanced flying machine ever built, part rocket, part airplane. Some 37 meters (122 feet) long, it has a wingspan of nearly 24 meters (78 feet). Empty, it weighs about 75 tonnes; ready for launch, over 90 tonnes. The orbiter is the only part of the shuttle system to go into orbit. It discards its booster rockets and external tank on the way. On its return from space it transforms from a spacecraft into a plane and uses its wings and tail to maneuver in the air and land on an ordinary runway.

The orbiter has three main engines (SSMEs, space shuttle main engines) mounted in a triangular configuration in the tail. They are mounted on gimbals, or swivel joints, so they can be swiveled to help steer the craft in powered flight. The SSMEs are the most powerful engines for their size ever built. When they fire together during lift-off, they develop a thrust of 500,000 kg (1.1 million pounds). It has been said that together, they produce enough power to light up the whole of New York State!

To develop such power, they have to consume liquid hydrogen and liquid oxygen at the incredible rate of 290,000 liters (63,000 gallons) per minute. They draw these propellants from the large external tank (ET) on which the orbiter rides into space. If they drew water from a family-sized swimming pool at the same rate, they would empty it in just 25 seconds!

Standing some 15 storeys high on the launch pad, the ET is the

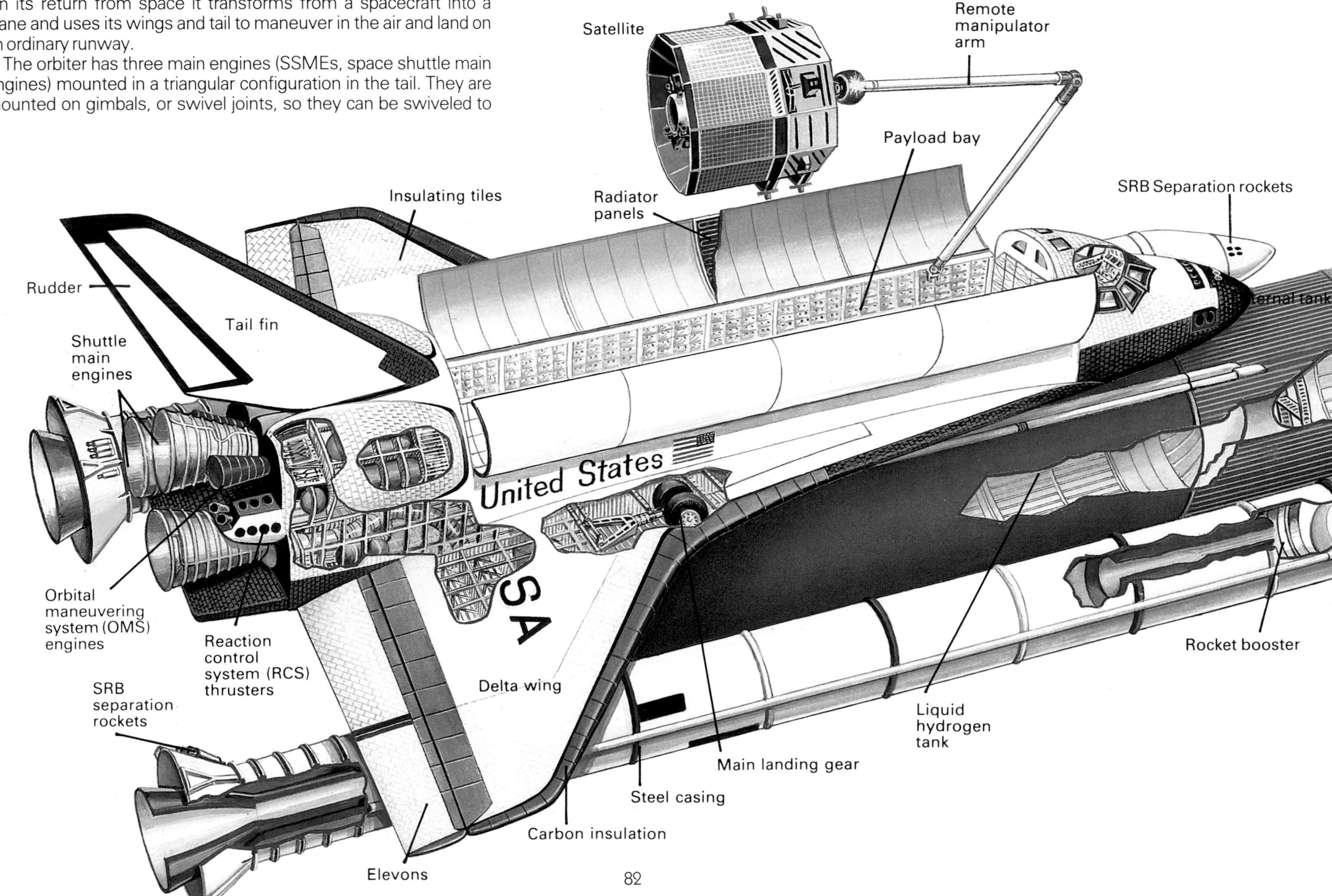

▼ **Shuttle system**
The external tank, holding propellants for the orbiter's main engines, provides the backbone for the shuttle stack. The orbiter is mounted on it, and the two solid rocket boosters are strapped to its sides. The orbiter and the boosters are all re-usable. Only the external tank cannot be used again. Most of the structure is made of aluminum alloys, except the booster rocket casings, which are made of steel.

▶ **CRT displays**
A multiple computer system holds the key to shuttle flight. The pilot and commander can call up all kinds of flight and engineering information from the system on to cathode-ray tube (CRT) displays in front of them in the orbiter's cockpit. In the picture part of the computer keyboards for commander and pilot are shown at bottom.

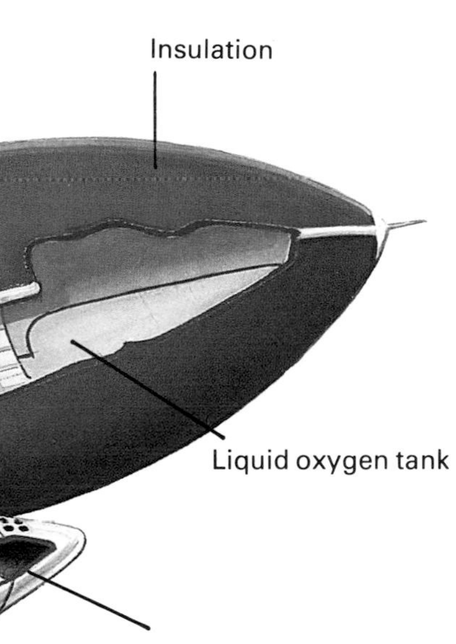

longest (47 meters, 154 feet) and, when loaded, the heaviest part of the shuttle hardware. It holds over 2 million liters (535,000 gallons) of liquid hydrogen and liquid oxygen propellants. It is the most colorful part of the shuttle, being orange-brown, which is the color of the insulation sprayed on it to protect the very cold propellants inside.

The powerful twins

Strapped to the sides of the ET on the pad are twin solid rocket boosters (SRBs), which provide the bulk of the thrust (about 80 per cent) at lift-off. They are the most powerful solid-propellant rockets ever built and the first to be used in manned space flight. The outer casing of the SRBs is built up by joining together a number of thick steel segments. The joints between the segments are sealed with sets of rings (O-rings), which prevent the escape of gases when the propellant burns. The failure of an O-ring and subsequent gas leakage was the cause of the *Challenger* disaster in 1986. Since then an extra O-ring has been added and other modifications carried out to prevent a recurrence of the problem.

The propellant itself is a mixture of fine aluminum powder (fuel) and ammonium perchlorate (oxidizer) in a synthetic rubber binder. When ignited, the SRBs each develop a thrust of over 1.4 million kg (3 million pounds). They fire at lift-off for about two minutes, before falling away and parachuting back to Earth to be recovered and used again.

Mounted alongside the main engines in the orbiter's tail pod are two smaller engines of the orbital maneuvering system (OMS). These engines are fired to inject the orbiter into orbit after main engine cut-off. They are also fired to change orbit in space, and as a retrobrake to bring about de-orbit. They burn hydrazine and

▲ **SRBs**
In the first stage of assembly of the shuttle stack inside the VAB, the SRBs are built up, segment by segment, on the mobile launch platform. Here assembly is nearly complete, with only the closeout of the joints to be carried out – a critical task.

▶ **Orbiter mating**
The orbiter is hoisted into the very top of the VAB before being lowered into position on the external tank, which is already mated with the twin SRBs.

nitrogen tetroxide propellants. These are hypergolic, which means that they burst into flame spontaneously when they mix.

The tail pod also incorporates two sets of thruster units, one on each side. There is another set of thrusters in the nose. Altogether there are 38 thrusters, which form the reaction control system (RCS). They are fired to provide attitude control in space. They burn the same propellants as the OMS engines.

Water power

Internal power for the orbiter is supplied by two systems – hydraulic and electrical. Hydraulic, or liquid-pressure, systems provide the power for operating the rudder and wing elevons and swiveling the engines. The pressure is supplied by pumps driven by small gas turbines, known as APUs (auxiliary power units).

The power for everything else, such as the lights, instruments, and computers, is provided by three power plants that produce electricity from three fuel cells. In a fuel cell, hydrogen and oxygen gases are made to combine chemically to form water and as they do so they produce electricity. The water they produce – about 3 kg (7 pounds) every hour – is very pure and is used by the crew for drinking and washing.

The airframe

Overall, the construction of the airframe is surprisingly conventional and essentially follows orthodox aircraft airframe technology. The structure is fabricated primarily in aluminum alloys, except in high stress areas like the tail, where higher-strength materials such as titanium are used. Since aluminum rapidly loses its strength at elevated temperatures, the airframe is covered with a variety of insulation, notably ceramic tiles (see page 92).

The crew ride, live and work in a pressurized cabin in the front fuselage of the orbiter. It is a three-level structure. The bottom level houses equipment. The middle deck area provides the main living accommodation (see page 114). And on top is the flight deck. At the front is the cockpit, where the commander (on the left) and pilot of each mission sit. At the rear of the flight deck is the payload station, where astronauts supervise payload operations (see page 127).

The cockpit looks much like the cockpit of a modern transatlantic airliner and has many of the airliner's flight instruments, such as artificial horizon and gyrocompass. But it has many more switches and buttons – 1400 in all – together with visual and audible alarms that flash and sound if things start to go wrong. In addition, the orbiter has three video display screens, which are linked to the on-board computer system. The astronauts can call up on the screen all kinds of flight data and information about any orbiter system.

Flying by computer

The shuttle's computer system is highly advanced. During critical phases of a mission, such as lift-off and re-entry, the system has to perform some 300,000 operations every second. Human astronauts, of course, would be totally incapable of coping with such demands and so take a back seat at such times. In fact, the computers fly the shuttle practically all the time – from a few minutes before lift-off to a few seconds before landing. And even when the astronauts do take over, they control the craft via the computer.

So critical is the computer system to shuttle operations that five separate computers are carried on-board – four main ones and a back-up. During a mission, the four main computers work continuously and process the same information. If one of the computers 'disagrees' with the others, it is overruled. If they all disagree, the back-up takes over.

The wisdom of multiple computer redundancy was demonstrated as early as November 1983 on STS-9, the first Spacelab flight. Two of the four computers were knocked out of action when the OMS engines fired to de-orbit prior to re-entry. NASA controllers feared that the others might follow suit. Fortunately they did not. With no computers at all, the orbiter, with its crew of six and $500 million payload, would certainly have been lost.

▼ **Rollout**
Mounted on the mobile launch platform the shuttle stack quits the VAB and begins its slow journey along the crawlerway to the launch pad. Carrying the stack is the eight-tracked crawler transporter, nicknamed the 'mighty tortoise'.

▲ **Crawlerway**
Flanked by swamps and lagoons on either side, the crawlerway to launch pad 39A leads in a straight line from the VAB. The crawlerway to pad 39B branches off to the right of the picture.

▼ **On the pad**
The launch platform is in position. The service structure will soon be moved around it.

▶ **Launch pad 39B**
The shuttle stack arrives at the launch pad. The crawler is slowly inching its way up the slope.

Assembling the stack

The preparations for a shuttle flight begin many months before launch day. The payload that will be carried into orbit has to be built and tested. The astronaut crew rehearse the operations that must be carried out during the mission. And the shuttle itself must be serviced, tested and assembled. The Kennedy Space Center (KSC) is the main shuttle launch site. A second site has been built at Vandenberg Air Force Base in California. It was scheduled to become operational in 1986, but was mothballed in the wake of the *Challenger* disaster. When it does eventually become operational, it will be used primarily by the military for launching classified payloads. It will also be used for civilian launches that need to be lifted into a polar orbit – one that goes over the North and South Poles.

At KSC the shuttle orbiter is serviced in the Orbiter Processing Facility (OPF). The payload is also often installed there. When processing of the orbiter is complete, it is towed next door to the mammoth Vehicle Assembly Building, the 'hanger' built to assemble the Saturn V Moon rockets in the Apollo era (see page 58). There the shuttle's SRBs are assembled and mated to the ET on top of a mobile launch platform. The tank is shipped by barge from a factory near New Orleans. Finally the orbiter itself is upended and secured to the ET. The shuttle stack is now complete.

When all is ready, the stack is rolled out to the launch pad by a gigantic eight-tracked crawler transporter, another relic of the Apollo era. It takes about six hours for the stack to reach the pad and for the launch platform to be secured. On the pad the stack is connected with external power and access arms from the tower-like service structure there. Routine tests of the shuttle take place at a relatively leisurely pace until the start of the countdown – the counting of time backwards to the moment of lift-off.

Countdown to lift-off

When the shuttle countdown begins, a huge digital clock is started at the press site at KSC, which shows how the countdown is proceeding. This clock does not show the actual time remaining before lift-off, however. This is because there are included in the countdown a number of 'built-in holds' – time periods when the clock is stopped.

Holds are built into the countdown to allow extra time in case the technicians preparing the shuttle fall behind with their work. Additional unscheduled holds may have to be made if any unexpected problems (such as bad weather) arise during the countdown.

The countdown for a shuttle mission typically starts about 56 hours before lift-off. But the countdown clock then shows, perhaps, as with STS-26, only 43 hours. The 13 hours difference is the time allotted for the built-in holds. These holds occur at specific times. For STS-26, for example, there were built-in holds at T-27 (T minus 27) hours, T-11 hours, T-6 hours, T-20 minutes and T-9 minutes.

As the countdown proceeds, the launch pad hums with activity. Power systems are hooked up, the fuel cells are fitted, the electric systems are tested and the communications links are checked out. In the Launch Control Center adjacent to the VAB, the consoles are brought into operation. At Houston in Texas, Mission Control

USA
United States
NASA
Challenger

◀ (far) **SRB separation**
About two minutes after the launch the SRBs run out of fuel and are cut loose. The orbiter's main engines continue firing for about six minutes longer. Then the external tank is jettisoned.

◀ **SRB splashdown**
Three enormous parachutes deploy above the SRBs at about 2000 meters (6500 feet) altitude and lower them into thc sea. There they float like buoys until they are recovered.

prepares to take charge of communications as soon as the shuttle leaves the pad. And at sites throughout the world tracking stations prepare to follow the shuttle when it enters orbit.

At about 'T-6 hours and counting' the final preparations on the pad begin. The external tank is filled with liquid hydrogen and oxygen. About four hours before launch, the flight crew are woken up, have breakfast and then kit-up. At about T-2 hours and counting, the crew climb into the orbiter. During the built-in hold at T-9 minutes, there is a final status check, and the launch director, in consultation with the mission management team (MMT), makes the decision to launch or no. If the decision is affirmative, the countdown resumes at T-9 minutes and counting.

The shuttle switches to automatic control from the ground launch sequencer: from this time on things happen too fast for human beings to be involved. At T-31 seconds the sequencer gives the go-ahead for the on-board computers to take over, and they complete the final automatic firing sequence.

Heads-down into orbit

At T-6.6 seconds, the orbiter's three main engines roar into life. The whole shuttle assembly rocks forward, but it is still firmly fixed to the mobile launch platform by eight hold-down bolts on the aft skirts of the SRBs. As the stack rocks back into the upright position, the SRBs fire. It is T-0. The hold-down bolts are then severed by initiators (small explosive charges), and the shuttle lifts-off.

After seven seconds the shuttle has cleared the tower and is heading for space. Operational control now switches to Mission Control, Houston. The crew, with their backs to the ground, are rammed by g-forces harder and harder into their seats as the shuttle gathers speed. Gradually the shuttle arcs over until the crew are in a 'heads-down' position.

The main engines give out transparent pale blue flames, rather like those on a gas cooker. The most luminous parts of the exhausts are little patches known as shock diamonds, created by shock waves in the gases. The SRBs on the other hand spew out thick columns of flame and smoke some 180 meters (600 feet) long. All the engines together develop a take-off thrust of nearly 3 million kg (7 million pounds).

Two minutes into the mission, and the shuttle has reached a height of about 50 km (30 miles). The SRBs are nearly out of fuel. Explosive bolts holding the SRBs to the fuel tank are then fired, and small rockets push them clear. They fall back to Earth. At an altitude of about 5 km (3 miles), they deploy three huge parachutes to slow them for a gentle splashdown in the sea, about 250 km (150 miles) down range. Recovery ships are on hand to pick them up and tow them back to KSC to be refurbished and used again.

Meanwhile, the remainder of the shuttle begins to accelerate more rapidly now that it is lighter and has left behind the densest part of the Earth's atmosphere. Some eight minutes after lift-off

◀ **Lift-off!**
The SRB's provide most of the flame and smoke, and over 80 per cent of the thrust, when the shuttle takes off. Clouds of steam envelop the launch platform as water is poured over it to keep it cool and absorb the shock waves created by the rocket exhausts.

◀ **Coasting in orbit**
Up in orbit, the orbiter coasts in a near circular path around the Earth at a speed approaching 28,000 km/h (17,500 mph). In the open payload bay of this orbiter (*Challenger*, STS-7) are pods holding satellites for launching. The pods protect the satellites from the uneven heating effects of the Sun while they are in the bay.

▶ (top) **Satellite deployment**
Some satellites are literally placed in orbit using the shuttle 'crane', or remote manipulator system (RMS) arm. The RMS plucks the satellite from the payload bay, moves it a safe distance away and then releases it. This gold-foil shrouded satellite, called ERBS (Earth radiation budget satellite), is deployed on STS 41-G.

▶ **Orbital maneuvering**
To change orbit up in space, the crew fire the twin engines of the orbital maneuvering system (OMS) in the tail pod of the orbiter. They will fire these engines again to de-orbit prior to re-entry.

the external tank begins to run dry and the main engines cut off. Explosive bolts connecting the orbiter with the tank are severed, and the tank is jettisoned. It gradually arcs over and drops back to Earth, smashing itself to pieces in the Indian Ocean.

About two minutes after tank separation the orbiter's OMS engines fire and inject the vehicle into an elliptical orbit. In only about a quarter of an hour, it has been boosted from rest to a speed approaching 28,000 km/h (17,500 mph). Thirty minutes later it has coasted to a height, typically, of about 250 km (150 miles). The OMS engines then fire again and place the orbiter into a more or less circular orbit. And it begins to coast round and round the Earth, making one revolution about every 90 minutes.

Orbital activities

The first priority of the crew when they enter orbit is to open the doors of the payload bay, the orbiter's cargo hold. This is vital to rid the vehicle of the heat generated by passage through the atmosphere and by all the electronic equipment on-board. The heat is radiated away from panels inside the doors.

A typical shuttle mission lasts for about five days, but may be longer or shorter according to requirements. Launching one or more satellites will generally be a major mission objective. With a length of 18.3 meters (60 feet) and a diameter of 4.6 meters, the payload bay can readily accommodate several normal size satellites at the same time. And, because the shuttle is manned, these satellites can be checked out in orbit to make sure they are working before they are abandoned. Other satellites can be serviced or repaired in space from the shuttle, or they can be brought back to Earth for repair.

Another major payload for the shuttle is the European-built Spacelab, a fully equipped space laboratory which remains inside the payload bay the whole time the shuttle is in orbit (see page 194). Spacelab missions are usually the longest to allow time for the experiments to be carried out. Other large payloads can be accommodated up to a maximum weight of about 23 tonnes. In the mid-to-late 1990s the shuttle will carry into orbit modules that will be assembled into the international space station *Freedom* (see page 200).

There is often spare space in the shuttle around its main payload. And NASA makes this space available for a modest fee to organizations or even individuals who would like to fly experiments in space. This 'getaway special' program makes it possible for much larger numbers of ordinary people to get involved in space-related activities.

Preparing for re-entry

At the end of their mission the orbiter crew close the payload-bay doors and make preparations to return to Earth. The next step is to de-orbit. By firing the RCS thrusters, the astronauts maneuver the orbiter until it is upside-down and traveling tail-first. They then fire the OMS engines for about two minutes. This has the effect of reducing the orbiter's speed to below orbital velocity, and allows gravity to pull it back to Earth. The RCS thrusters fire again and jockey the orbiter into a nose-up position with its underside facing forwards.

About 45 minutes after the de-orbit burn, the orbiter slams into the outer fringes of the atmosphere at an altitude of about 120 km

(75 miles), traveling at 25 times the speed of sound. As the air grows thicker, it progressively arrests the orbiter's headlong plunge. Friction between the air and the underside of the orbiter makes this glow red-hot, with temperatures soaring to over 1300°C.

Unprotected, the aluminum airframe of the orbiter would melt away at this temperature, and the astronauts inside would be burned to a frazzle. But the outer skin of the orbiter is made up of highly efficient heat-resistant materials. The underside is covered with thick, black-coated tiles made of silica fibers. Much of the upper surface of the wings, tail and fuselage is covered with thinner white tiles. In all, more than 30,000 tiles are used, each individually contoured for its specific location on the airframe. The nose and the leading edges of the wings, which experience the highest temperatures, are covered with a highly refractory reinforced carbon material. Less vulnerable areas, such as the upper fuselage and payload-bay doors, are covered with flexible felt insulation. Safe inside the orbiter, the crew see a pinkish glow from the red-hot tiles light up the windows.

As the air around the orbiter heats up during re-entry, it ionizes, or becomes electrically charged. The presence of this ionized layer prevents the passage of radio waves, which means that the crew cannot communicate with, or receive communications from, Mission Control. They are in this communications 'blackout' for approximately 15 minutes, an anxious time for both them and Mission Control.

Heading for touchdown

By the time the orbiter emerges from the communications blackout, it has slowed down to about seven times the speed of sound. The air is rapidly thickening, and the aerodynamic surfaces of the orbiter begin to bite. The elevons on the wings, the speed brakes and rudder on the tail take over attitude control from the RCS thrusters.

As the orbiter glides towards the landing site, it performs a series of roll reversals that help slow it down. It finally loops round and heads for the runway. It descends seven times more steeply than an ordinary airliner. Just before touchdown, the pilot brings up the orbiter's nose and lowers the landing gear. The typical landing

▼ Touchdown
The orbiter lands on a runway like an ordinary airplane but somewhat faster, at a speed approaching 350 km/h (250 mph). It needs a much longer runway than a commercial airplane because it can't brake by reverse thrust like jet planes do.

◀ **Safing**
Immediately the orbiter lands, a ground support crew moves in to make the vehicle safe. They remove any residual fuel in the OMS and RCS tanks and purge systems of toxic and explosive gases. Only then is it safe for the crew to leave.

NASA
905
United States
NASA
Challenger

▲ **Kennedy landing**
Within sight of the twin launch sites of pads 39A and 39B, *Challenger* makes the first landing at the Kennedy Space Center, on 11 February 1984, at the end of mission 41-B.

◀ **The 747 carrier jet**
When the shuttle lands at a site away from Kennedy, it has to be returned there on a modified Boeing 747 carrier jet. Most landings now take place at Edwards Air Force Base in California, where the runway stretches for many kilometers across the desert.

speed is about 350 km/h (220 mph). On touchdown the astronauts apply the wheel brakes, and the orbiter rolls to a halt after about 2500 meters (8000 feet). They do not leave the craft immediately. They emerge only after it has been 'safed' by a ground crew to prevent the risk of fire from explosive gases inadvertently leaking from the thruster systems.

The landing sites

In theory the prime shuttle landing site is the 5-km (3-mile) runway at KSC, a few kilometers from the launch pads. This became operational in February 1984 when *Challenger* touched down on completion of STS 41-B. With the exception of STS-3 (March 1982), all previous flights had landed at Edwards Air Force Base in the Mojave Desert in California, where there is plenty of room for the orbiter to overshoot in an emergency. (STS-3 landed at White Sands Missile Range in New Mexico when Edwards AFB became waterlogged.)

However, the notoriously unpredictable weather at the Cape – which can cause totally different conditions over the length of the runway – frequently compromises landing safety. And until forecasting problems there have been resolved, shuttles will routinely land at Edwards AFB. The KSC site is, of course, still operational for emergency use in flight abort situations if anomalies occur during the launch phase. There are several other alternative emergency landing sites dotted around the world, which are manned by NASA personnel during missions. They include Zaragoza and Moron Air Bases in Spain, Ben Guerir in Morocco and Banjul in the Gambia, West Africa. Where these sites have comparatively short runways, NASA has installed huge nylon nets to snare the orbiter if it overshoots. It is also evaluating the use of braking parachutes (which the Soviet shuttles use) to reduce the landing distance.

When the orbiter lands at Edwards AFB or an alternative landing site, it is transported back to KSC by a specially modified Boeing 747 carrier jet. It is then towed into the OPF, where it is thoroughly overhauled and prepared for its next mission. Many of the tiles may have to be replaced, as may some of the bigger units, even the engines. But it will not be long before the orbiter is lifting off the pad with another astronaut crew and heading once more into space.

Back to the future

The date is Thursday, 29 September 1988; the location, launch pad 39B at the Kennedy Space Center. On the pad is orbiter *Discovery*, mated to its boosters and external tank and waiting for the off for its seventh and the shuttle's 26th mission (STS-26). Inside, a veteran astronaut crew is poised to attempt the most significant American space flight since the space debut of the shuttle in April 1981.

It is 32 months since *Discovery*'s sister orbiter *Challenger* left that same pad, only to be blasted into oblivion 73 seconds into the mission (see page 132).

The time is 11.26 am; and the countdown is in a built-in hold. Launch supremo Robert Crippen, pilot of the first shuttle flight, consults his mission management team and then decides. It's Go for launch. At 11.28 the countdown is resumed at T-9 minutes. With seconds to go, *Discovery*'s three main engines burst into life. Then at T-0 the twin SRBs belch flame and the shuttle is cut loose. It rises swiftly from the pad, and over the loudspeakers at the Space Center the voice of Launch Control Hugh Harris announces triumphantly to the throngs of spectators: 'Americans return to space, as *Discovery* clears the tower!'

But the joy is muted as time ticks away towards the 70 second mark. Capcom at Mission Control, Houston, now in control of the flight, informs the crew: 'Go at throttle up.' Commander Rick Hauck on *Discovery* acknowledges: 'Roger go.' Everyone watching cannot help but remember that it was at this moment in the previous shuttle flight on 28 January 1986 that *Challenger*

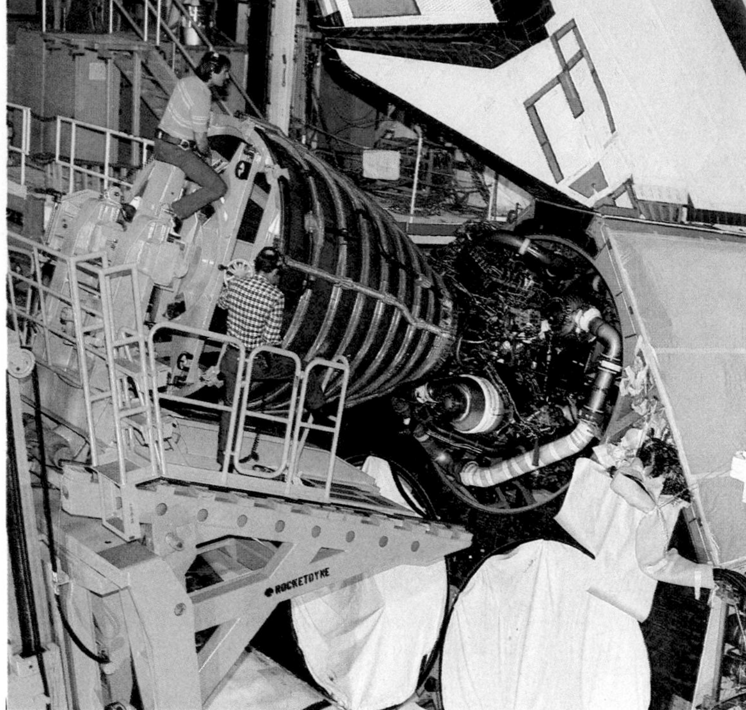

◀ **STS-26 SRB assembly**
Exceptional care is taken in the assembly of the modified solid rocket boosters in the build-up to the launch. Here the aft section of one booster is ready to be lifted on to the launch platform.

▲ **STS-26: Main engines**
One of the three main engines being installed in *Discovery*. Thirty modifications have been made to the design to improve reliability.

exploded. Not until *Discovery* jettisons the SRBs two minutes into the flight do the real cheers begin, rising to a crescendo when *Discovery* safely enters orbit six and a half minutes later.

NASA Administrator Dr James Fletcher, for one, is beside himself with excitement. 'Wow!' he exclaims, 'That was really something. It's been a long wait. It's the first of a new era.'

In Washington the news is relayed to President Reagan, who announces at the beginning of an awards ceremony: 'America is back in space. I think I had my fingers crossed like everybody else.' Later, in inimitable prose, *Time* magazine summed up the mood of America at that moment: 'The nation's collective sigh of relief could have launched a thousand shuttles.'

The new beginning

Discovery remained in space until Monday 3 October, during which time the crew successfully deployed a tracking and data relay satellite (TDRS) to replace the one lost on *Challenger*. They also carried out a series of experiments, one of which (protein crystal growth) sought to grow superior crystals of an enzyme that triggers AIDS.

▶ **STS-26: Escape pole**
A Navy parachutist tries out the new shuttle escape system fitted to *Discovery* for STS-26, the first post-*Challenger* shuttle flight.

Discovery dropped out of the sky after its 64th orbit of the Earth. It signaled its return to California with a double sonic boom, which drew cheers from the crowd waiting at Edward Air Force Base. As it touched down, Mission Control at Houston observed: 'A great ending to the new beginning.' When *Discovery*'s crew emerged an hour later, George Bush was there to greet them in his capacity as Vice-President (but not unmindful of the fact that he was currently running for President!). Another VIP present was the legendary Chuck Yeager, the first person to fly through the sound barrier.

A new bird

As the STS-26 shuttle stack sat on the launch pad waiting for lift-off on 29 September, it looked no different from the stack that carried *Challenger* on its last fateful flight. But that outward appearance was deceptive. In fact every major system in the whole shuttle system – orbiter, SRBs and ET – had been thoroughly checked over the intervening months with regard to their efficiency, reliability and safety. As a consequence more than 400 major and minor modifications were made.

Over 150 changes were made to the SRBs, root cause of the *Challenger* disaster. A third O-ring and a metal seal were incorporated into the joints between the segments that make up the rockets. One hundred extra bolts were added to the nozzle area. The solid propellant itself was recontoured internally to reduce stress when it burns. Eight changes were made to the ET, including a redesign of the valves that control propellant flow to the orbiter's main engines. And the main engines themselves underwent over 30 changes. The engine-pod structure and combustion chamber were strengthened. The blades and bearings of the turbopumps, which feed propellants into the combustion chamber, were upgraded to improve their durability.

It was *Discovery*, though, which underwent the bulk of the modifications – some 220 in all. Improvements were made to the RCS and OMS engines and to the APUs. The heat shield was upgraded in certain areas. The brakes, tires and steering gear were beefed up to absorb the wear and tear of heavy high-speed landings.

As far as crew safety was concerned, the astronauts flew into orbit kitted out in new orange partial-pressure suits. Each suit had its own oxygen tank, parachute and inflatable life-raft. *Discovery* itself had fitted a new emergency escape system for use during a launch abort or after re-entry, when the orbiter was in controlled gliding flight. The escape system consisted of a 3.6-meter (12-foot) telescopic pole, which the astronauts would extend after blowing off an escape hatch. Then they would, in turn, fix a ring on their suit to the pole and slide along and out into the air, deploying a parachute and drifting gently back to Earth.

On that STS-26 launch day the five crew of *Discovery* were well aware of their responsibilities as they prepared to return America to manned space flight in a space vehicle that, as one commentator

1

put it, 'may as well be a new orbiter on its maiden flight'. The flight was also a test for NASA's new management team. Said *Discovery*'s pilot Richard Covey: 'The thing we don't forget as a crew is that a lot of these changes have not been flight-tested yet, and that our mission is, indeed, a flight-test of those changes.'

Discovery and its crew came through these tests triumphantly. As they punched their way through the clear blue Florida sky on 29 September 1988 the American dream was being restored. Headlined *Time* ecstatically: 'The Magic is Back!'

A Soviet *Snowstorm*

It was no coincidence, of course, that the Soviets chose the launch day of *Discovery* on mission STS-26 to reveal to the world for the first time its own shuttle craft. Named *Buran* (*Snowstorm*), it is at first sight a US space shuttle orbiter lookalike. Explained Novosti Press Agency's Mikail Chernyshov: 'The similarity has been prompted by laws of aerodynamics.' That said, *Buran* has obviously benefited hugely from the freely available data on its US counterpart. However, there are as we shall see significant differences between the two reusable craft.

According to Moscow, the development of a Soviet shuttle began in 1978, while the US was still struggling to get its own shuttle act together. Ten years and some $10 billion later, it was ready for its maiden flight. The first launch attempt was announced for 29 October 1988. It was to be a remote-controlled unmanned flight, blasting off from, and touching down at, the premier Soviet spaceport, the Baikonur Cosmodrome.

But with just 51 seconds of the countdown remaining, the launch computer detected 'an unconventional situation' and shut down the launch sequence. It had detected that the crew access arm, which is left in position until the last moment in case of an emergency, had not retracted sufficiently, and there was danger of the launch vehicle hitting it on lift-off. The launch was postponed until the error in the system had been corrected. Not until 15 November was another launch attempted, but this time there were no glitches. *Buran* rose out of the giant fireball that enveloped the pad and disappeared into the heavens. Eight minutes later it separated from its booster rocket, firing on-board engines to accelerate it finally into orbit 250 km (150 miles) high. After two orbits of the Earth, *Buran* fired its engines again to de-orbit. Three hours 25 minutes after lift-off, it made a perfect landing just 12 km (7.5 miles) away from the launch site.

Eulogized Radio Moscow: 'The space plane has ushered in a new era in the history of Soviet space exploration.' The Soviets in fact had a double cause for celebration because only a few days earlier cosmonauts Vladimir Titov and Musa Manarov in space station *Mir* had broken the world space endurance record of 326 days (see page 200).

A different bird

Like the US orbiter, *Buran* is a delta-winged craft the size of a medium-range airliner. It has broadly similar dimensions to the US craft. Its length is 36.4 meters (119 feet), a fraction shorter than the US orbiter. Its wingspan is more or less identical, at 24 meters (79 feet). At 16.5 meters (54 feet) tall, *Buran* is not quite the height of the US orbiter, while its diameter, at 5.6 meters (18.3 feet), is a little greater. The take-off weight is about 10 tonnes more, at over 100 tonnes.

Compared with the US orbiter, *Buran* has one very major difference. It does not have main engines to help boost it into orbit (the US orbiter has three). Instead, *Buran* rides into space piggy-back on the world's most powerful expendable rocket, Energia (see page 74). This behemoth of a rocket, 60 meters (197 feet) long, puts out 170 million horsepower as it thunders into space. It made its first successful flight on 15 May 1988, carrying a dummy payload.

◀ (far) **STS-26: Lift-off!**
Their help not needed this time, an emergency escape crew watch *Discovery* soar into the heavens on 29 September 1988 at the beginning of a thankfully flawless mission.

◀ ***Discovery*'s crew are feted**
on their return to the Space Coast at October's end. Messages flash at them from the motels. This one occupies a special place in space history since it was where the 'Original Seven' astronauts stayed in the 1960s when they pioneered US manned space flight.

▼ **STS-26: Welcome home**
Vice-President George Bush is waiting at Edwards Air Force Base on 3 October 1988 to welcome home the happy crew after their four-day mission. The modified SRBs put in a flawless performance. From the front, the crew are Frederick Hauck, Richard Covey, Mike Lounge, David Hilmers and George Nelson.

▸ **Tail end**
Soviet technicians move in to check and make safe the on-board engines and thrusters in the tail section after *Buran*'s successful first flight. This took it twice around the Earth, and lasted for nearly 3½ hours.

Another major difference in the operation of the Soviet shuttle is that it can be carried out completely automatically, as was demonstrated with *Buran*'s maiden voyage. Undoubtedly the trickiest part of the automated mission is the landing. Any slight deviation from the nominal flight regime or instrument or systems failure at this stage must result in loss of the vehicle. One of the designated Soviet shuttle cosmonauts, Igor Volk, was said to have opposed an unmanned first flight for this reason, arguing that in the event of systems failure, a cosmonaut could fly in the craft manually. It is thought that Volk and fellow cosmonaut Ural Sultanov could be the crew for the first manned flight of *Buran*, probably in 1989 after another unmanned verification flight.

Buran design

Like its American counterpart, *Buran* is made up of three main sections. The nose houses a pressurized crew cabin. This was unoccupied on its maiden flight, but is big enough to accommodate a cosmonaut crew of up to four, together with six 'passengers'. The 'passengers' would probably be what the Soviets call cosmonaut researchers, the equivalent of NASA's payload specialists (see page 105).

The mid-section of *Buran* is a payload bay with opening doors, measuring 4.7 meters (15 feet) across and 18.3 meters (60 feet) long, comparable with that of the US shuttle orbiter. In the bay it can carry payloads of up to 30 tonnes. This is considerably in excess of what the US orbiter can currently manage (about 23 tonnes). The aft section of *Buran* comprises the tail pod, in which are housed orbital maneuvering engines for injecting the craft into orbit at launch and for effecting the de-orbit burn prior to re-entry.

During re-entry *Buran* uses aerodynamic drag to slow down. As a heat shield it has some 38,000 ceramic tiles, black underneath

▾ **Off to launch**
Buran is carried out to the launch pad horizontally on a transporter that travels on twin rail tracks. The locomotives that haul the transporter can be seen here. On the pad *Buran* is upended by means of the machinery at the aft end of the transporter.

and white on top like the US orbiter. It maneuvers in its gliding path through the atmosphere by means of elevons on the wings and a rudder on the tail. When it has landed on the 4.5-km (3-mile) long runway at Baikonur, it deploys drogue parachutes to help reduce its speed quickly. The 'chutes are jettisoned once its speed falls to 50 km/h (30 mph).

The VKK fleet

Buran is the first operational member of the Soviet shuttle orbiter fleet, which already includes another two craft at different states of readiness. The next to fly will probably be one called *Ptichka*, or *Birdie*. The orbiters form part of the launch system which the Soviets call VKK – Vozdushno-Kosmichesky Korabl (air-spacecraft). This is the equivalent to the American STS (Space Transportation System), the technical name for the space shuttle system. To handle VKK launches elaborate new facilities have been built at Baikonur, in addition to the runway already mentioned. These include two huge workshops, the assembly-and-test buildings, where the orbiter/Energia stack is put together. Unlike the US shuttle, the stack is assembled in the horizontal position, orbiter uppermost. It is then rolled out on a transporter, hauled by four diesel locomotives on a twin rail track. The machinery on this mobile transporter-erector helps turn the stack into a vertical position on the pad.

▼ On the pad
Ready for its maiden flight on 15 November 1988, *Buran* is mounted on the immensely powerful Energia rocket. This launch configuration invites interesting comparison with that of the US space shuttle. Of the launch hardware, only *Buran* is reusable.

Chapter 4

LIVING IN SPACE

◀ **Smile please!**
A happy shuttle crew poses for the camera during the maiden flight of the third operational orbiter, *Discovery*. The six-day mission, 41-D, launched on 30 August 1984, sees the second American woman in orbit, Judith Resnik. Gathered around her (clockwise from top right) are Steven Hawley, Michael Coats, Henry Hartsfield, Michael Mullane and Charles Walker.

▶ **Raspberry ripple**
Mike Lounge chases a bubble of raspberry drink during a meal break on the four-day STS-26 mission that lifted off on 29 September 1988. The crew are eating in the mid-deck area of *Discovery*. Looking on are commander Frederick Hauck (left) and David Hilmers. Note the family snapshots stuck to the lockers, which adds a homely touch.

Many of the fears voiced at the beginning of manned space flight, querying whether flesh and blood human beings could physically withstand the rigors of space flight, have been allayed. Properly equipped and protected, astronauts can tolerate, even revel in, the space environment. Soviet cosmonauts have put this beyond dispute with their year-long sojourns in space station *Mir*.

The days when all astronauts had to be fine physical specimens, test pilots with nerves of steel and possessing the 'right stuff' are also numbered. In the modern generation spacecraft like the space shuttle the astronaut crew experience relatively mild g-forces on launch and re-entry. And they live in orbit for the most part in a pressurized, shirt-sleeve environment, akin to that in a high-flying airliner. So ordinary people are now able to go into space with the minimum of training.

But we must not be lulled by the increasing frequency and apparent ordinariness of space flight into complacency. Space remains a deadly frontier. This was underlined by the explosion that blasted *Challenger* out of the Florida sky in January 1986. The *Challenger* disaster, said one commentator, served as 'a salutary reminder of the enduring hostility of space and the fragility of man'.

This chapter looks at the workaday lives of the astronauts up in orbit and the training they undergo before they get there. Out-of-the-spacecraft, or extravehicular, activity is covered in the next chapter.

MF71C
MF57E
AUTOMATED DIRECTIONAL
SOLIDIFICATION FURNACE

OVER THE DECADES since the beginning of the Space Age, the criteria on which astronauts are selected have changed radically. This was inevitable really because in the 1950s nobody knew what space was like and how, or indeed whether, man would fit into it. In early spacecraft such as Mercury, the human astronaut was, to be honest, not really necessary. His job could be done by instruments or monkeys, and, on several occasions, was.

According to an early training manual issued by McDonnell Douglas, manufacturers of the Mercury capsule, the capsule crew 'consists of one man representing the peak of physical and mental activity, training and mission indoctrination.... [He] must not only observe, control and comment upon the capsule system, but must scientifically observe and comment upon his own reaction while in a new strange environment.'

Eventually NASA announced that astronauts (all male) would be selected on the basis of seven criteria: (1) Age less than 40 years; (2) Excellent physical shape; (3) Height less than 5 feet 11 inches; (4) A degree in engineering, or equivalent; (5) Qualified jet pilot; (6) Graduate of a test-pilot school; and (7) With a minimum of 1500 hours flying experience. On the basis of these criteria and President Eisenhower's stipulation that they must be military personnel, the 'Original Seven' astronauts were selected for Project Mercury (see page 38).

The 5 foot 11 inch limit on height was dictated by the size of the Mercury capsule, whose size in turn was dictated by that of the available launch vehicles. But even an astronaut of this height found the capsule a very tight squeeze. As Apollo 11 veteran Michael Collins has so graphically put it, the Mercury astronaut was 'bundled up like a mummy wrapped inside a tight sarcophagus'.

The professionals

The new breed of astronaut has no such rigid requirements imposed upon him, or her (space is no longer the male chauvinist realm that it once was). True, the astronauts who actually fly the

◀ Crew training
Four of the five STS-26 crew take up their launch positions for a practice countdown in the crew compartment trainer at the Johnson Space Center, Houston. The trainer is outfitted exactly like the real thing. This view looks towards the rear of the flight deck. Note the windows that look forward into the payload bay, and upward into (in orbit) space.

▶ Soyuz training
Soviet cosmonaut Viktor Gorbatko (left) and 'guest' cosmonaut Pham Tuan from Vietnam are pictured training in a Soyuz capsule at the Yuri Gagarin Cosmonaut Training Center at Star City, near Moscow. The two will ride into space together on the Soyuz 37 mission in July 1980, heading for rendezvous with the Salyut 6 space station.

spacecraft into orbit have to be more like the astronauts of yesteryear. They are skillful jet pilots with considerable flying experience. On the shuttle two such pilot-astronauts fly on each mission, commander and pilot, and they occupy the forward seats in the orbiter cockpit.

Another category of professional astronauts on the shuttle are termed mission specialists. They are not required to know how to fly the orbiter. Their job is to ensure that the mission objectives are met. They are instrumental in launching satellites and looking after other payloads the shuttle may carry. They operate the remote manipulator system (RMS) – the shuttle's crane. And they are trained for spacewalking.

The mission specialists are in general qualified scientists or engineers. So are the third class of astronauts aboard the shuttle, the payload specialists. They are not career astronauts but mainly scientists who have a particular interest or expertise in an experimental payload being carried, and they are responsible for its operation. Sometimes a payload specialist is included for a slightly different reason. NASA sometimes offers a place aboard to countries whose satellites are being launched. For example, Prince Sultan Abdul Aziz Al-Saud was a payload specialist on STS 51-G (June 1985) to participate in the launch of the Arabsat communications satellite from *Discovery*. In general a payload specalist goes into space only the once.

On shuttle mission 51-D in April 1985, *Discovery* had a typical mixed-specialist crew. The two pilot-astronauts were Karol Bobko (commander) and Donald Williams (pilot). Stanley Griggs, Jeffrey Hoffman and Rhea Seddon were mission specialists. Griggs and Hoffman were called upon to perform an unscheduled EVA (the first in the shuttle program) when the Leasat-3 satellite malfunctioned immediately after launch. Seddon helped them by operating the RMS. Payload-specialist Charles Walker's responsibility was to operate an electrophoresis experiment to separate biological samples. The other payload specialist, Senator Jake Garn, was really more of a guest, though he did take part as a guinea pig in medical experiments.

▼ **Mission simulator**
The most realistic 'hands-on' training of shuttle astronauts takes place in the shuttle mission simulator at Johnson. The instruments are 'live' and respond to the pilot's actions as they would in real flight. Note the hydraulic rams on which the mock cabin is mounted. These deliver the muscle to move the cabin this way and that to provide added realism.

▶ Flying training
The pilot-astronauts spend quite a lot of their time flying in specially modified jets like the T-38A. This jet has modified speed brakes fitted – you can see them below the mid-fuselage. Their purpose is to alter the jet's aerodynamic performance on the landing approach so that it corresponds with that of a gliding shuttle orbiter.

▶ High gs
This is the centrifuge at the Johnson Space Center, in which the astronauts get a foretaste of the high g-forces they will experience when they go into, and return from, space. They ride in the cabin, or gondola, which is whirled around on the arm at high speed.

The typical crew on a Soviet Soyuz flight includes commander, flight engineer and cosmonaut researcher. The latter two are the equivalent of mission or payload specialists, depending on the mission. Through a program called Intercosmos, Soyuz may carry a 'guest' as a cosmonaut researcher. The 'guests' come mainly from other socialist states, such as Cuba and Vietnam, and also from other countries, such as India. The Soviets also have close links with France in the space field and have included the Frenchman Jean-Loup Chretien in two crews – in June 1982 (Soyuz T-6) and, most recently, in November 1988 (Soyuz TM-7) visiting space station *Mir*. Unusually for a 'visiting' cosmonaut, Chretien went spacewalking on this last mission.

Astronaut training

Astronauts are selected for particular flights months in advance of the launch date. And they undergo appropriate training for their individual roles and also general training to ready themselves for life in space. The astronauts' general training includes preparing for the weightlessness they will experience when they climb in orbit. They get brief snatches of weightlessness in arcing aircraft, and longer periods suited up in water tanks. Tank training is practiced most widely by mission specialists who are scheduled to go spacewalking (see page 136). All the crew also take part in emergency escape sessions on land and water, and on the launch pad.

The pilot-astronauts spend a good deal of time flying in specially modified jets and finely tuning their flight and navigation profiles.

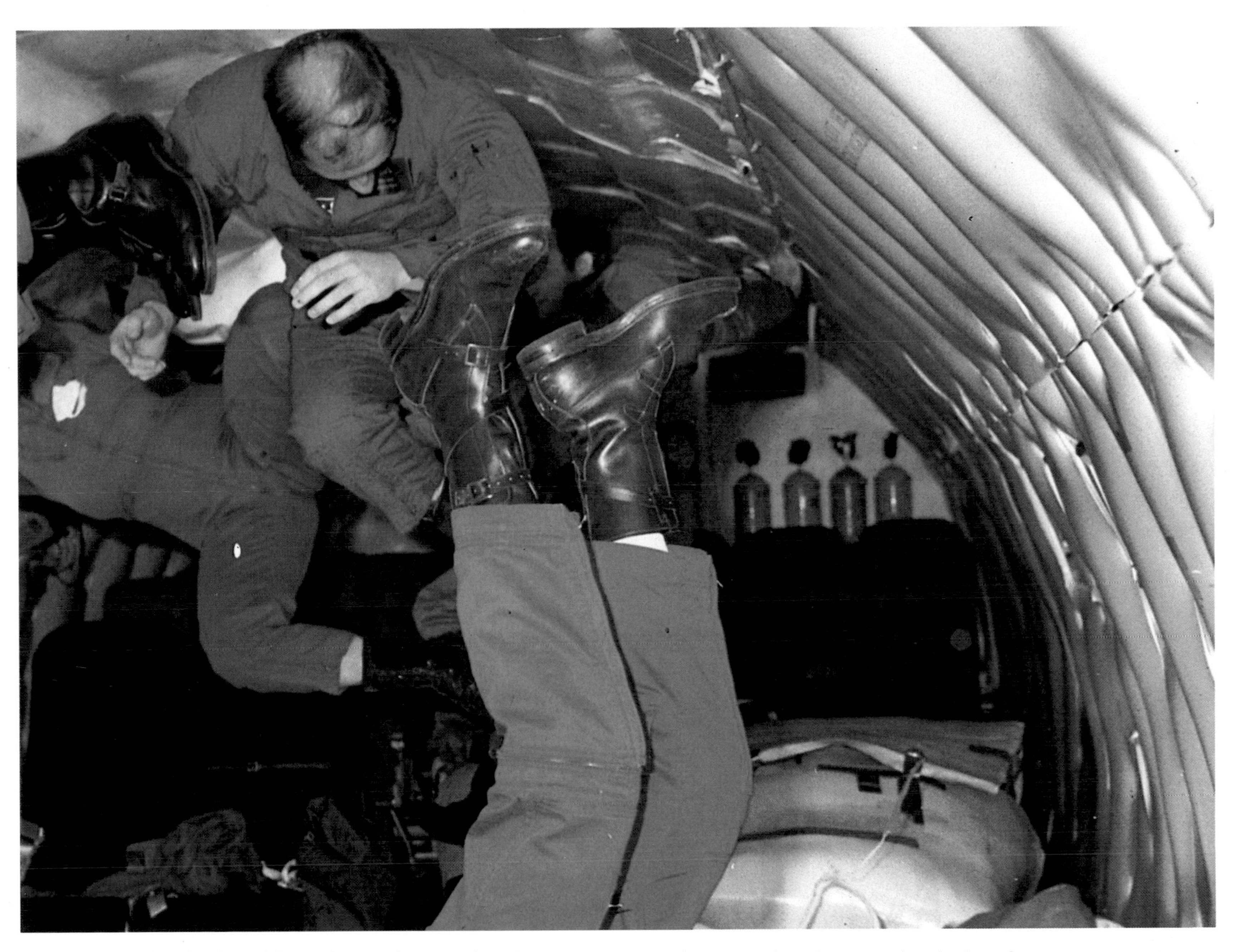

▲ Low gs
Training astronauts experience a few seconds of weightlessness in a KC-135 aircraft. It provides fleeting zero-gravity conditions as it flies up and over in a tight arcing trajectory. Note the padding on the walls.

The mission and payload specialists rehearse the operations they will carry out in orbit. All categories of astronauts train a lot in mock-ups and simulators. Mock-ups are full-scale representations of spacecraft, in whole and in part. They are virtually identical to the real thing and enable the crew to familiarize themselves with the layout of instruments and controls.

The most realistic training takes place in simulators, mock-ups with the added ingredient of live instruments. In these computer-controlled machines the instruments respond in the same way as the real ones would when the switches and controls are activated. An example of the ultimate in this technology is the shuttle mission simulator at Johnson Space Center. This features a 'live' mock-up orbiter cabin mounted on hydraulic legs, which tilt the cabin in response to thruster and engine control activation. Computer-generated pictures are projected on screens to show appropriate views through the windshields. The crew go through detailed countdown and lift-off simulations in conjunction with Mission Control as the launch date draws near.

The major part of American astronaut training is conducted at the Johnson Space Center, Houston, home of Mission Control. Additional training takes place at the Marshall Space Flight Center at Huntsville, Alabama. Soviet cosmonaut training is concentrated in a suburb of Moscow called Zvezdnyy Gorodok, or Star City. There is located the Yuri Gagarin Cosmonaut Training Center, which has similar facilities to Johnson, including mock-ups, simulators, centrifuges and laboratories. The Center has seen the

◀ **Soyuz water training**
Soviet cosmonauts flying Soyuz ferry craft return to Earth inside the re-entry module. They aim to touch down on land, but must be prepared for a watery landing if things go wrong.

training of all Soviet cosmonauts from Gagarin on and, by the end of 1988, nearly 70 had made it into orbit.

The weightless world

From the launch pad the journey into space takes a surprisingly short time, only about a quarter of an hour. The crew always travel into space with their backs towards the engines. In this way their bodies can withstand the g-forces that build up as their launch vehicle accelerates from rest to the orbital velocity of about 28,000 km/h (17,500 mph). For shuttle travelers the g-forces scarcely exceed 3g, which is about the maximum level terrestrial joyriders experience on the latest type of fairground roller coasters. The early astronauts experienced some 6-8g on lift-off.

On the journey up into orbit the crew feel a certain amount of vibration and buffeting as they punch their way through the atmosphere. But as the air thins out, the flight becomes smoother. After the final thrust of the engines to insert them into orbit, the vibration and the noise ceases. And they begin effortlessly circling the Earth. Viewing the terrestrial landscape unfolding gradually below, there is little sensation of speed. And it is difficult for the crew to grasp that they are actually traveling at 25 times the speed of sound.

When the astronauts unstrap themselves from their seats, they begin to drift about. Gravity appears to be absent. But it isn't of course. The astronauts are actually falling around the Earth, but they don't feel this because their spacecraft and everything else is falling with them.

Endless somersaults

Weightlessness dominates everything that the astronauts try to do in orbit, moving, eating, drinking and going to the bathroom. For astronauts with a gymnastic bent, space is a wonderful place. An endless somersault can be performed effortlessly – in fact it can be initiated involuntarily by the simple act of bending down. Walking is not possible because there is nothing to keep the feet down on the floor – not that 'floor' (implying down) has any real meaning in space. Astronauts move about by pushing and pulling themselves along. But they must remember that, although weightless, they have inertia, that is, the tendency to carry on moving when they try to stop themselves.

The absence of a stable floor, ceiling and walls not surprisingly has a disturbing effect on the body. The brain initially can't cope because the eyes and the organs of balance in the ears send confusing messages to it. In more than a third of the astronauts and cosmonauts that venture into space, such messages result in bad attacks of motion sickness. The astronauts sweat, feel extremely nauseous and are frequently physically sick.

Fortunately, most recover from space sickness, or space

▼ Rehearsing experiments
Here Ulf Merbold (right) from Germany fixes electrodes on Dutch 'guinea-pig' Wubbo Ockels when conducting body-function monitoring exercises. Merbold and Ockels are payload specialists in training for Spacelab missions.

▶ Quick exit
M113 armored personnel carriers like this are located near the shuttle launch pad. In a pad emergency the crew would make for the carrier, batten down the hatches, and beat a hasty retreat.

adaptation syndrome, after a day or so. But on many flights it can seriously jeopardize the crew's efficiency. Strangely, many of the early astronauts, in Project Mercury for example, never succumbed to space sickness. This was probably due to the fact that they were traveling in capsules that permitted them very little movement and enabled their brains to cope more easily. For the record cosmonaut Gherman Titov holds the dubious distinction of being the first man to be spacesick, on the second manned spaceflight, Vostok 2, in August 1961. (He was also the first person, incidentally, to go to sleep in space.)

Haute cuisine

From the astronauts' point of view one of the worst aspects of the early space flights was the food. Having braved untold dangers and risking life and limb every second they were aloft, they were

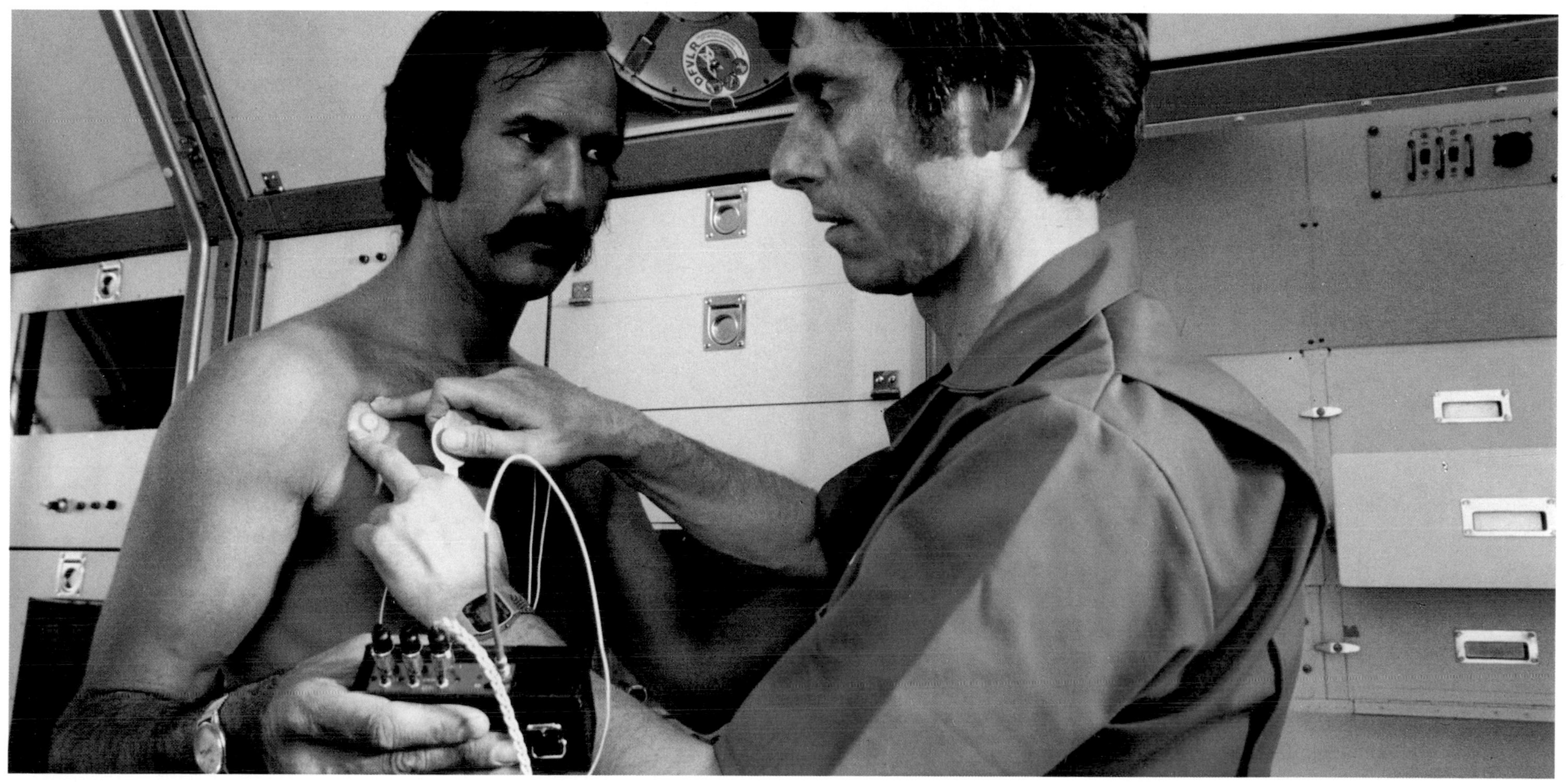

Discovery
United States

◀ **Lift-off!**
The spectacle of a shuttle launch will never be captured on film more dramatically than this. This photograph of *Discovery* on 24 January 1985 (STS 51-C) is taken with an IMAX camera, shooting with 70 mm film and a wide-angle lens. The camera is located in a specially engineered firebox at the 78-meter (255-foot) level of the launch gantry.

▶ **Fun with custard!**
Joseph Allen conducts an impromptu experiment to determine the effects of zero gravity on a custard dessert. He maneuvers it out of the container (top) and notices it forms a spherical globule. Now comes the tricky part, getting it back in again. But, yes. Allen succeeds at the first attempt (bottom). Joy upon joy!

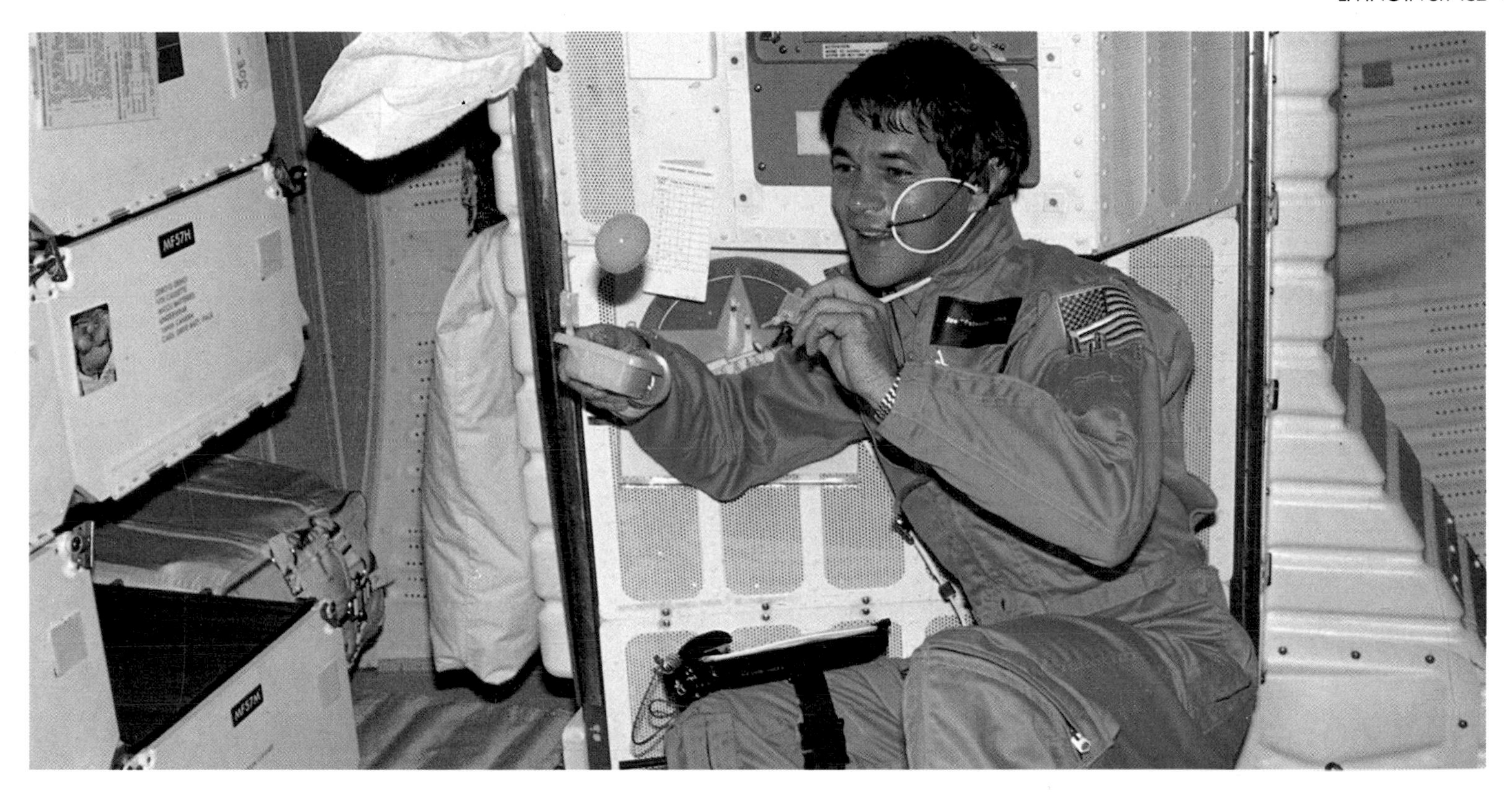

presented at mealtimes with a squirt of gooey paste from a tube! It bore little resemblance to real food even though it contained the requisite nutrients and calories.

Gemini astronaut Virgil Grissom was particularly known for his abhorrence of space-food goo, which is what prompted the notorious corned-beef sandwich incident on the first Gemini flight (see page 51). NASA was not amused, and this led to Associate Administrator Dr Mueller solemnly assuring Congress that NASA had taken steps 'to prevent a recurrence of corned-beef sandwiches' in future flights!

Fortunately for today's astronauts, tasteless goo is a thing of the past. Shuttle astronauts, for example, dine remarkably well, having the choice on a week-long mission of some 70 dishes. And each crew member can select his meal individually. On day three of the STS-26 mission in September 1988, pilot Richard Covey had three meals, which included: for breakfast scrambled eggs, bran flakes and orange-mango drink; for lunch chicken salad, fruit cocktail and almonds; and for dinner Teriyaki chicken, creamed spinach, candy-coated peanuts and vanilla pudding. This not only gave him a balanced diet of 3000 plus calories for the day, it also stimulated his taste buds.

Much of the food taken on-board is dehydrated and has to be mixed with water before eating. Some comes in cans or pouches and some comes in natural form. Sandwiches come in bite-size packs and are treated with gelatine to prevent the release of crumbs. In weightlessness the crumbs would rapidly spread into the cabin and maybe get into delicate instrumentation.

The meals are prepared by one of the astronauts at the galley unit, which includes a water dispenser for rehydration and an oven. He or she assembles the food and drink packages on a tray, from

▶ Topsy-turvy world
Posing for a crew portrait on Earth is never like this! In the weightless environment of space the astronauts can assume and hold any position they choose. The question is, who is upside-down? Commander of the crew on this *Challenger* mission (51-F) is Gordon Fullerton in the striped shirt. The others are, in clockwise order, Anthony England, Karl Henize, Roy Bridges, Loren Acton and John-David Bartoe, with shaven-headed Story Musgrave in the middle. Henize, at the age of 58, is the oldest man ever to fly in space.

◀ Hair-raising
Making her space debut on STS 41-D is the second US spacewoman Judy Resnik. In the absence of gravity her luxuriant tresses expand to enormous proportions. Note the pad suspended in mid-air to her right.

◀ (below) **Slinky**
Taking a break on STS 51-D, Jeffrey Hoffman and Rhea Seddon play'n learn with a slinky toy. They are trying to determine what effect zero gravity has on its operation. This comes as a welcome relief from their efforts to activate the satellite Leasat 3, which failed to function after deployment.

which the crew eat – it is rather like an in-flight meal on an airliner. The diner uses a knife, fork and spoon in the usual terrestrial manner. Because the foods are in the main moist, they stick to fork and spoon as long as these are not moved too quickly.

A czar's banquet

Soviet space food tends to be less adventurous than that served on the space shuttle. But on the Soyuz TM-7 mission to *Mir* in November/December 1988 all this changed, at least temporarily. The driving force behind the change was the visiting cosmonaut, Jean-Loup Chretien. He had flown with Soviet cosmonauts six years previously and had protested then about the unappetizing gooey pastes he was served, which offended his Gallic palate.

So on the latest mission he took with him what one of his Soviet colleagues called a 'czar's banquet'. It included rabbit with prunes, duck with artichokes, and cabbage stuffed with dates and raisins. This exotic fare must have been particularly welcomed by two of the cosmonauts on *Mir*, who had been in space for a year. Explaining the importance of this *haute cuisine spatiale,* one of the chefs who prepared it, Pierre Rouage, said: 'Cosmonauts are deprived of everything, even carnal pleasures, so they need something like a good meal to remind them of home.'

EOS

Suck it and see

Drinking presents a problem in zero-g. The astronauts can't pour themselves a drink because liquids don't pour. Pouring needs gravity. They can't drink from glasses for the same reason. However, they can drink liquids by sucking through a straw because this method relies on air pressure rather than gravity, and their spacecraft is pressurized, of course. There is no shortage of water on the space shuttle because it is produced in abundance (3 kg, 7 pounds an hour) by the fuel cells that provide electrical power.

Running water is available on the shuttle in the hand washbasin in the side of the galley unit. The 'running' is achieved by means of a flow of air since water can only run properly under gravity. For a good all-over wash, the astronauts must content themselves with a sponge-down with a soapy cloth, followed by one to rinse. Only a little water is needed because in zero-g water clings well to the skin. Splashing too much water about in zero-g will send little droplets everywhere.

On the Skylab space station the crews had the luxury of a shower. When one of the astronauts showered, he used a water gun to squirt warm water over his body. But his two colleagues had to stand by to vacuum up the water drops escaping from the shower compartment! On the shuttle anyway there is no room for a shower. But one is planned for the living quarters of space station *Freedom* (see page 200).

Hygiene and sanitation in general are more important in a spacecraft than in the home on Earth. The reason is that in the close confines of a spacecraft, in zero-g, germs can multiply rapidly. So illness could spread among the crew like wildfire. For this reason 'cleanliness is next to Godliness' at all times. There is no washing of dishes as such because all the food containers are thrown into plastic trash bags. Eating utensils and trays are wiped down with wet wipes; living areas are regularly wiped over with disinfectant.

Each astronaut has a personal hygiene kit that includes toothbrush, toothpaste, dental floss, soap, comb, clippers, antichap lipstick, skin lotion and deodorant stick. The male astronaut's kit in addition includes a safety razor and shaving cream. Wet shaving is the rule, for the beard whiskers stick to the cream and can be removed with a cloth. Mechanical shavers would scatter the whiskers into the air.

Among the other personal items astronauts may have are sunglasses, scissors (essential for opening food packages), a Swiss army knife, a chronograph and a supply of Velcro, that ingenious hook-and-eye plastic fastener. The astronauts use this to attach things to a surface to prevent them floating about.

Digestive elimination

The phenomenon of weightlessness has several unpleasant

◀ (far) **Bon appetit!**
This still from the IMAX film 'The dream is alive' shows the STS 41-D crew tucking into one of the three square meals they enjoy each day. Compared with the early astronauts, they eat in a most civilized manner. From left to right, the diners are Henry Hartsfield, Judy Resnik (almost hidden), Michael Mullane, demonstrating a dynamic method of eating with a fork, Charles Walker and Stephen Hawley.

◀ **The gooey stuff**
Don't be fooled by the vodka labels, the tubes actually contain food! Apollo-Soyuz astronaut Donald Slayton has been given the tube-food by his Soviet colleagues on the mission. He does not seem to be very impressed with it!

aspects, not the least of which is going to the lavatory – a process NASA terms digestive elimination. If one pauses to think about it, this biologically frequent activity does pose a problem in zero-g. Muscles in the body function perfectly normally and discharge the wastes from the body. But what then? The urine forms into a stream of hovering droplets, and the feces just hang around!

In the early days of space travel, the astronauts had to make do, rather messily, with plastic bags containing germicidal chemicals. Now these bags are carried for emergency use only. Although their proper name is emesis bags, they are always called 'doggie bags' by the astronauts!

The shuttle astronauts use for their digestive elimination a passable imitation of the terrestrial water closet. But the Space-Age closet is flushed by air, not water. And it has other features that set it apart from its Earth-bound relative – foot restraints, hand holds and a seat belt!

The closet is a commode-like device which has a separate hose

▼ Coke is it!
Anthony England is one of the first to enjoy a Coke in space on STS 51-F. Wags term this development 'a big burp for man'!

▶ Exercising the muscles
STS-2 commander Joe Engle keeps in trim by running on the spot in this elasticated exercise harness. On later shuttle flights a proper treadmill will be fitted for in-flight PT.

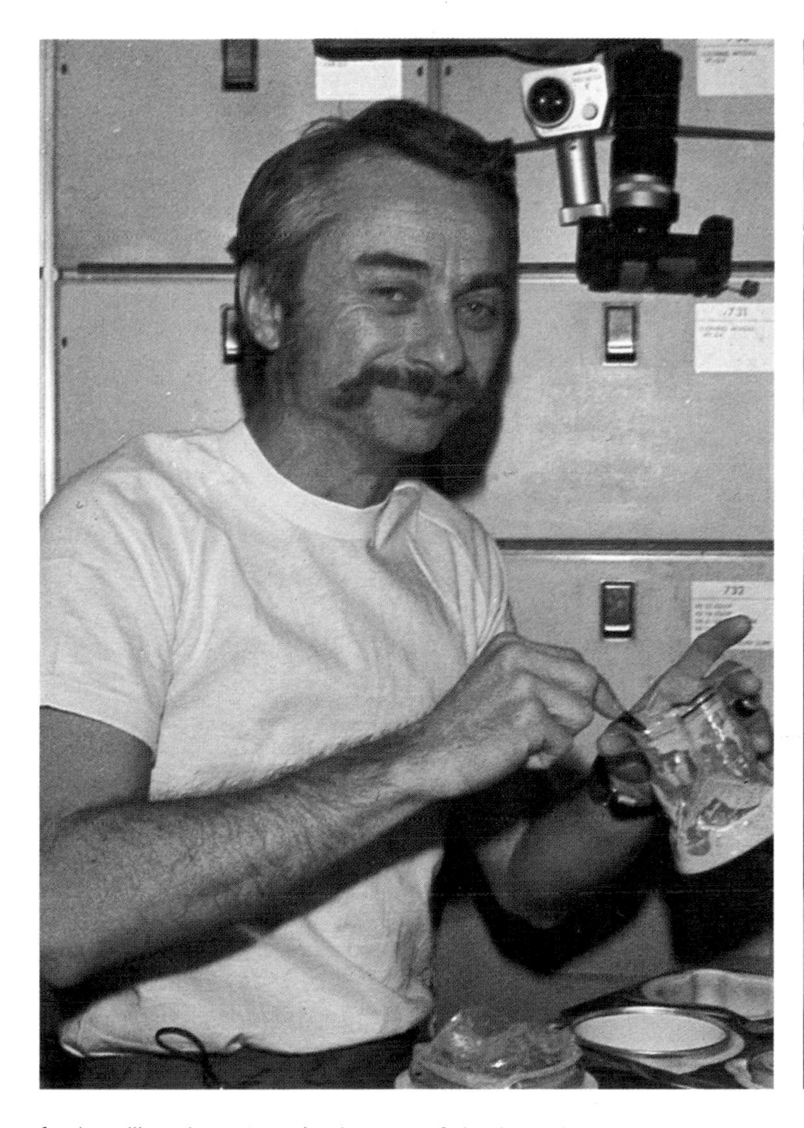

for handling the urine. At the top of the hose is a contoured cup designed for use by both sexes. The urine is drawn into the on-board waste-water tank and from time to time is dumped overboard. Feces are drawn down into a lower chamber and shredded by a kind of fan called a slinger, which flings them on to the chamber walls. There they are dried by exposure to space vacuum and disinfected. But they are not dumped overboard.

And so to bed

There is no such thing as going to bed at night on an orbiting spacecraft. The astronauts see the Sun rise and set about every 90 minutes! So they have a sleeping period imposed on them, typically 7-8 hours in every 24 hours. Because of zero gravity, conventional beds and bed clothes are useless. The astronauts sleep zipped up in sleeping bags, or sleep restraints. The restraints need to be anchored to something to prevent them drifting away.

In the shuttle orbiter sleep stations are located on the mid-deck. They take the form of horizontal bunks, and the astronauts can attach their restraints to either side of the bunks, though some prefer to 'hang vertically' on the nearby lockers. In each case they tie themselves down firmly so that their body makes contact with a hard surface. This has proved to be the best method of fooling the body that it is sleeping on a normal bed, and a more restful sleep ensues.

When sleeping, most of the astronauts make use of the standard-issue eyemasks and earplugs. This is not really surprising. A spacecraft is never totally dark and there is always some background noise, as fans whirr, motors hum and switches click automatically on and off. And, of course, the emission of a variety of noises by up to eight astronauts gathered together in one place is not inconsiderable!

Life support

Of all the systems on-board a spacecraft, none is more important than life support. This provides the means of keeping the astronauts alive while they are in orbit. There are many aspects of life support, including the provision of food and water and the disposal of wastes, which we have already considered. But the most vital one is the environmental control system, which provides

▲ (left) **Skylab fare**
On the second Skylab mission in 1973 Owen Garriott enjoys a meal. Skylab astronauts are the first to eat with ordinary cutlery. They eat at a table in the living compartment.

▲ **A trim sir?**
William Lenoir trims Robert Overmyer's sideburns on *Columbia*'s fifth flight in a row in November 1982. The astronauts carry all the necessary gear for keeping neat and tidy in their personal hygiene kits. One is shown stuck to the locker above Lenoir's head.

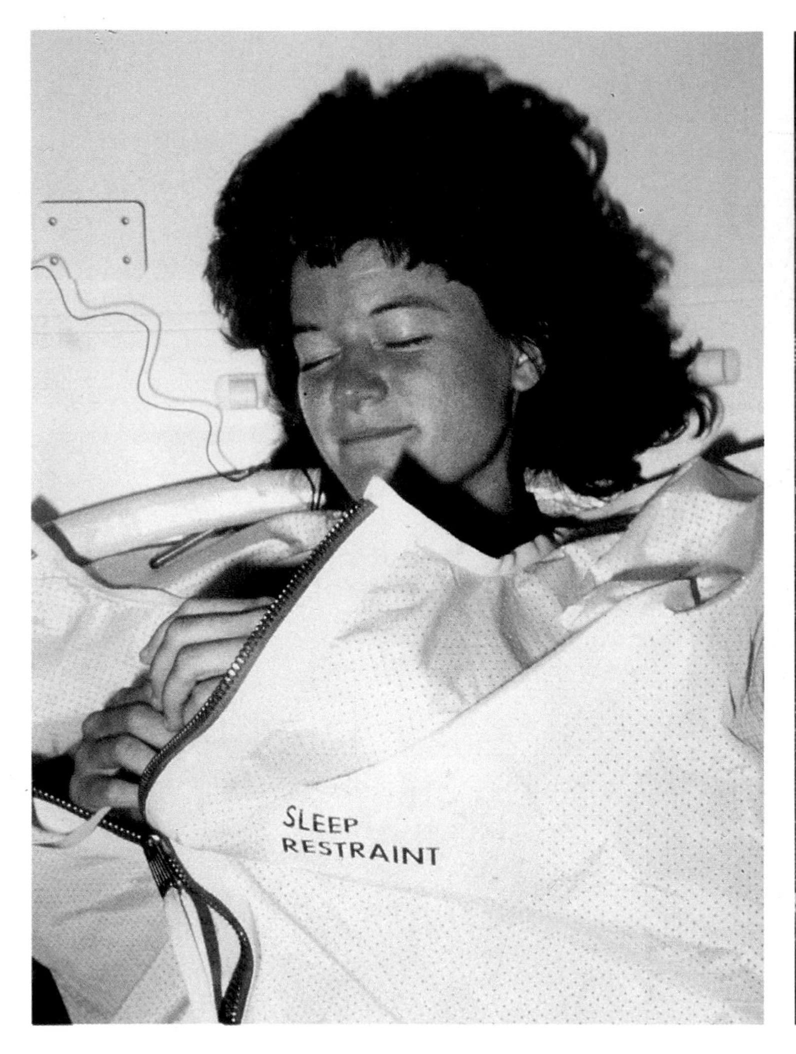

Restrained sleep
To prevent themselves floating away from their beds, astronauts sleep zipped inside sleeping bags, or sleep restraints as they are appropriately termed. Arms can optionally be zipped in, as demonstrated by Sally Ride, or left out.

▶ (far) **Kaliningrad control**
At all times during manned space flight, Mission Control is in contact with the spacecraft, checking telemetry from its instruments and tracking it around the Earth. This is the Mission Control Center for Soviet manned missions. It is located in the city of Kaliningrad on the Baltic Sea coast, due west of Moscow.

a breathable life-sustaining atmosphere for the astronaut crew.

On Earth we human beings are used to breathing air at atmospheric pressure, I kilogram per square centimeter (14.7 pounds per square inch). The higher we go above sea level, the thinner the air becomes, or in other words the lower is the air pressure. By 3000 meters (10,000 feet) human beings begin to get breathless and may develop headaches. By 7500 meters (25,000 feet) they are starved of oxygen and their breathing is very labored; gases may start to come out of the blood, giving them an attack of the bends. Above 9000 meters (30,000 feet) they are unconscious through lack of oxygen. At about 19,000 meters (63,000 feet) their blood starts to boil. Above 80 km (50 miles) there is still a little air but essentially we are in space. Spacecraft like the shuttle fly in an orbit about 250 km (150 miles) high. Outside in space at that altitude, there is a vacuum far more perfect than anything that can be attained on Earth.

This vacuum of course is deadly to human life, which has to be cocooned inside a secure pressure cabin, supplied with a breathable atmosphere. Modern spacecraft, such as the shuttle and the Soviet Soyuz ferries and space stations, use an 80/20 mix of nitrogen and oxygen gases, which approximates to the composition of ordinary air. The pressure, too, is atmospheric. This contrasts with the atmosphere used in the Apollo spacecraft, which was pure oxygen at one-third of an atmosphere pressure.

The environmental control system in a spacecraft is essentially a super air-conditioning unit. It not only supplies the oxygen and nitrogen, but also controls the temperature, pressure and humidity of the atmosphere. It removes exhaled carbon dioxide by circulating the air through canisters of lithium hydroxide, which absorb the gas. It uses filters to remove dust and lint. And it uses activated charcoal to extract odors, which would otherwise quickly build up to an intolerable level in a confined space that is at the same time workroom, kitchen, dining room and bedroom for a crew of up to eight.

Rays from the cosmos

The atmosphere on Earth not only gives us oxygen to breathe, but also plays other vital roles in preserving life. For example, it protects our planet from extremes of temperature between day and night. During the day the air helps absorb the heat of the Sun and distribute it more evenly. At night it acts like a blanket to prevent too much heat from escaping back into space.

In airless space, however, temperatures are extreme. In the sunlight temperatures can soar sky high. This is why spacecraft

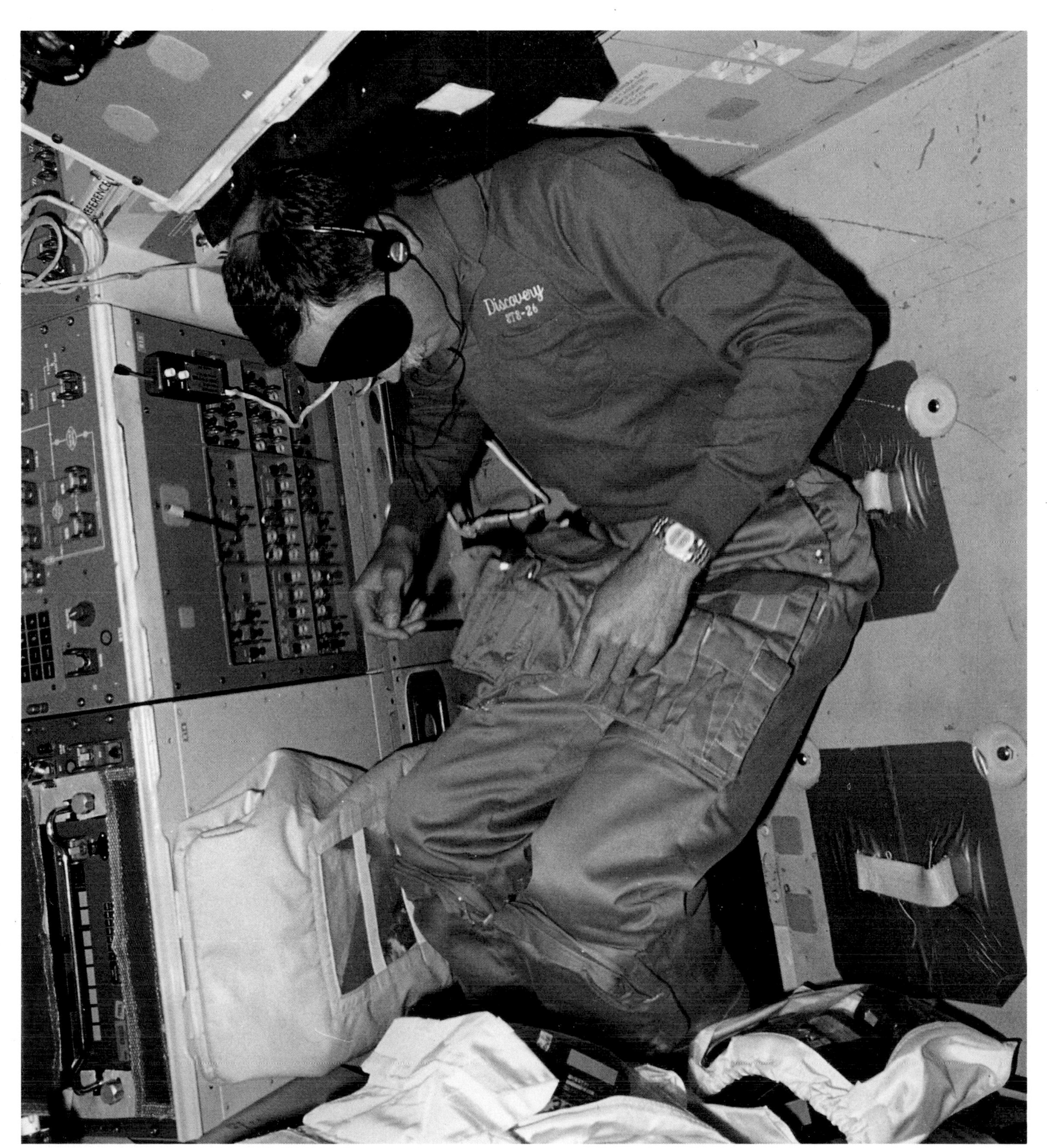

◀ **Catching 40 winks**
Pilot Richard Covey catnaps on the aft flight deck of *Discovery* during the STS-26 mission. Sensibly he has Velcroed himself to the deck to prevent himself floating away. His relaxed body has assumed the typical fetal position that a body does in zero-g.

USA

◀ **Orbiter in orbit**
The first thing the shuttle pilot does when reaching orbit is open the doors of the payload bay. This exposes radiators on the inside of the doors. They dump the excess heat that has built up in the orbiter during its climb into orbit. On this mission, STS-7, *Challenger* is making its second flight into orbit.

▼ **Checking the systems**
For most of the time in orbit the spacecraft functions in automatic mode. But the commander and pilot regularly check over their spacecraft's systems and update their flight plan. If anomalies occur at any time in the systems, however, the on-board computers trigger flashing lights and audible alarms to alert the crew.

components are usually painted white or are covered with gold foil to reflect much of the Sun's heat. They are often set spinning so that no one side gets overheated. Out of the sunlight, however, temperatures plummet. Any heat a body possesses rapidly radiates away.

The Sun poses another hazard. It gives out a constant stream of charged particles, which we call the solar wind. Astronauts in a spacecraft must be protected from such particle radiation, which can damage human tissue, affect the genes and, in large doses, kill. More random radiation coming from outer space generally, the so-called cosmic radiaton, is also hazardous. Fortunately, the metal that forms the pressure shell in a spacecraft crew cabin is thick enough to block most of the radiation.

Some cosmic rays, however, do penetrate the cabin. And it is for this reason that astronauts wear a dosimeter at all times. This is a device that registers the amount of radiation he or she receives. Cosmic rays also manifest themselves visually to the astronauts. They create flashes of light on the retina even when the eyes are closed. The radiation problem is tolerable in Earth orbit, but it will become much more serious when man ventures forth on interplanetary missions (see page 204).

Deadly debris

Another natural hazard in orbit is presented by meteoroids, or tiny bits of rock. Such bits frequently rain down on the Earth, causing the streaks of light we call meteors. The increasing amount of man-made debris that is presently accumulating in orbit poses an even greater hazard.

Experts believe that pea-size fragments of rock or debris hitting a spacecraft at the orbital speed of some 28,000 km/h (17,500 mph) could punch a hole clean through it. And there are estimated to be anything up to 70,000 such fragments of that size or larger in orbit

▼ Debris damage
This pit is found in one of the windshields of *Challenger* after it has returned from its second trip into space in June 1983 (STS-7). The damage is thought to have been caused by man-made debris because the pit contains traces of titanium oxide, which is a pigment used in spacecraft paint.

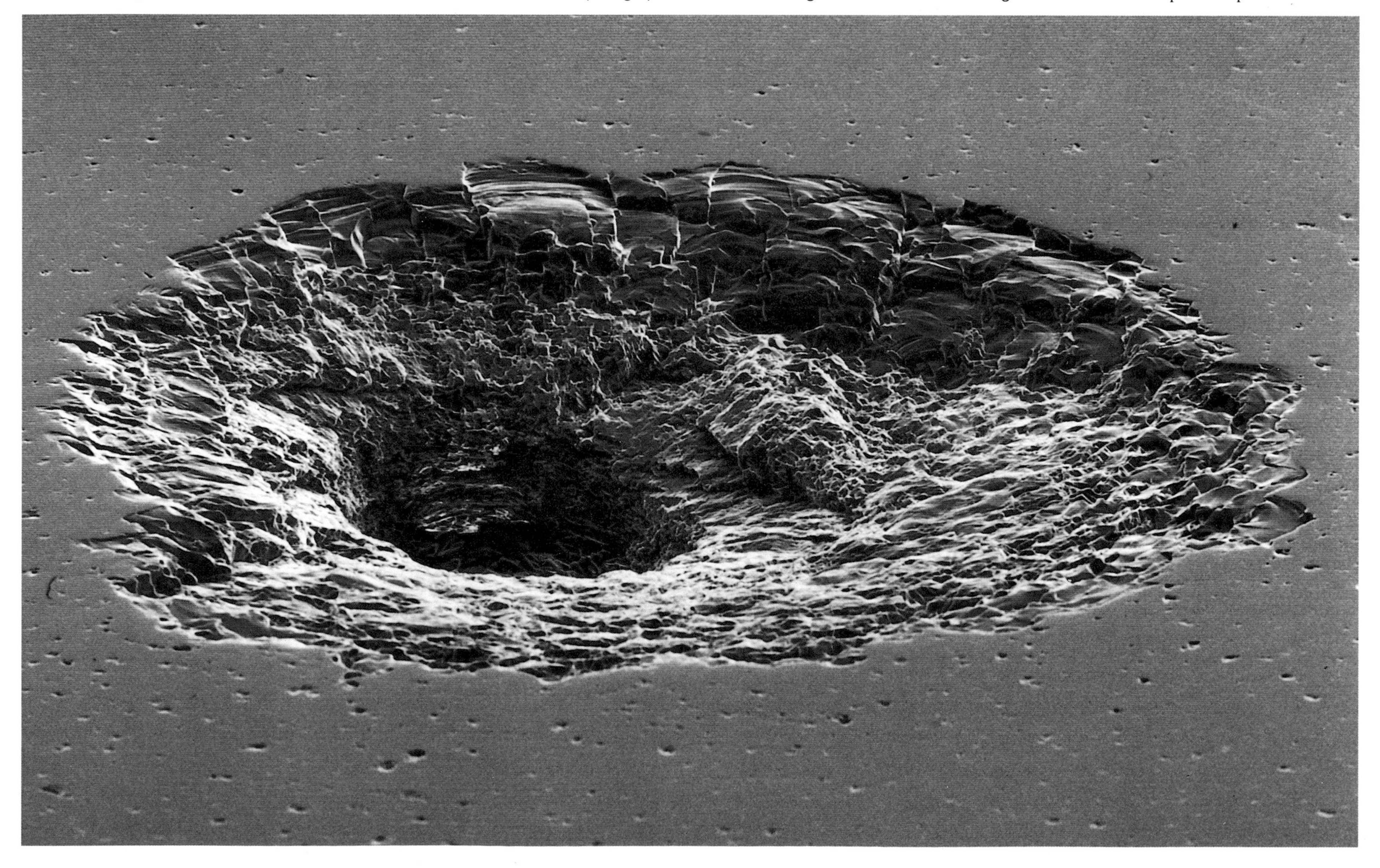

TDRS-C deployment
One of the most important shuttle payloads springs out of *Discovery*'s payload bay, on the first post-*Challenger* mission, STS-26, in September 1988. It is the communications satellite TDRS (the tracking and data relay satellite). It is heading for geostationary orbit, where it will join TDRS-A. The folded umbrella-like structure on top will later deploy into a dish antenna.

◀ **LDEF deployment**
With the distinctive Baja California as a backdrop, the 9-meter (30-foot) long satellite known as the LDEF (long-duration exposure facility) is deployed by *Challenger*'s robot arm on STS 41-C. Carrying a host of experiments, it is scheduled to be recovered in a year's time.

▶ **Launching with PAM**
Launching satellites is a major shuttle activity. Some are launched by being sprung from a pod in the payload bay. To this satellite, Anik C-3, is attached a rocket stage that will later fire to boost it up to geostationary orbit. The stage is known as PAM (payload assist module).

▶ (far) **Aft flight station**
Pilot Robert Overmyer pictured at the in-orbit flight station at the rear of *Columbia*'s flight deck. Behind him is one of the main flight controls, the orbiter rotational hand controller. It controls the firing of thruster jets to rotate the craft. Note Overmyer's plump face. This is a consequence, not of over indulgence in shuttle cuisine, but of the redistribution of his blood in zero-g.

at the present time! One of the windshields of *Challenger* was hit by a small piece of debris on the STS-7 mission in June 1983 and had to be replaced. Only a month later the cosmonauts orbiting in the space station Salyut 7 heard a crack and discovered a 4-mm diameter crater in one of their windows.

Mission operations

On the shuttle a primary task on most flights is to launch satellites. Because the payload bay is so large – some 18 meters (60 feet) long and 4.6 meters (15 feet) across – it can accommodate more than one satellite at a time. As early as mission STS-7, the orbiter was launching three satellites at once. On that occasion these were: Anik C-2, a comsat for Canada; Palapa B, a comsat for Indonesia; and a satellite called SPAS (shuttle pallet satellite). This latter satellite was used to practice deployment and retrieval maneuvers with the RMS arm. It carried a camera which snapped the best pictures ever taken of a shuttle orbiter (in this case *Challenger* in orbit (see page 120).

The two comsats were carried into space in protective insulated cradles in the payload bay. The insulation was necessary to prevent their contents overheating in the Sun. For the launch they were first set spinning for stabilization, then they were sprung from the cradle. To the base of each was attached a rocket motor called PAM (payload assist module), which was later fired to boost the satellite into geostationary orbit.

The frisbee launch

A different method of satellite deployment is used for larger satellites. They are rolled out of the payload bay sideways in what is termed frisbee style. The first satellite to be launched in this way was Leasat-2 on STS 41-D, orbiter *Discovery's* maiden flight in August 1984. This was one of three successful comsat launches

▶ Hurricane!
At a safe distance in orbit the crew of STS 51-I chart the inexorable progress of hurricane Elena in September 1985. On the ground Elena is wreaking great devastation. The calm eye in the hurricane center brings an all-too-brief respite from winds spiraling at speeds in excess of 160 km/h (100 mph).

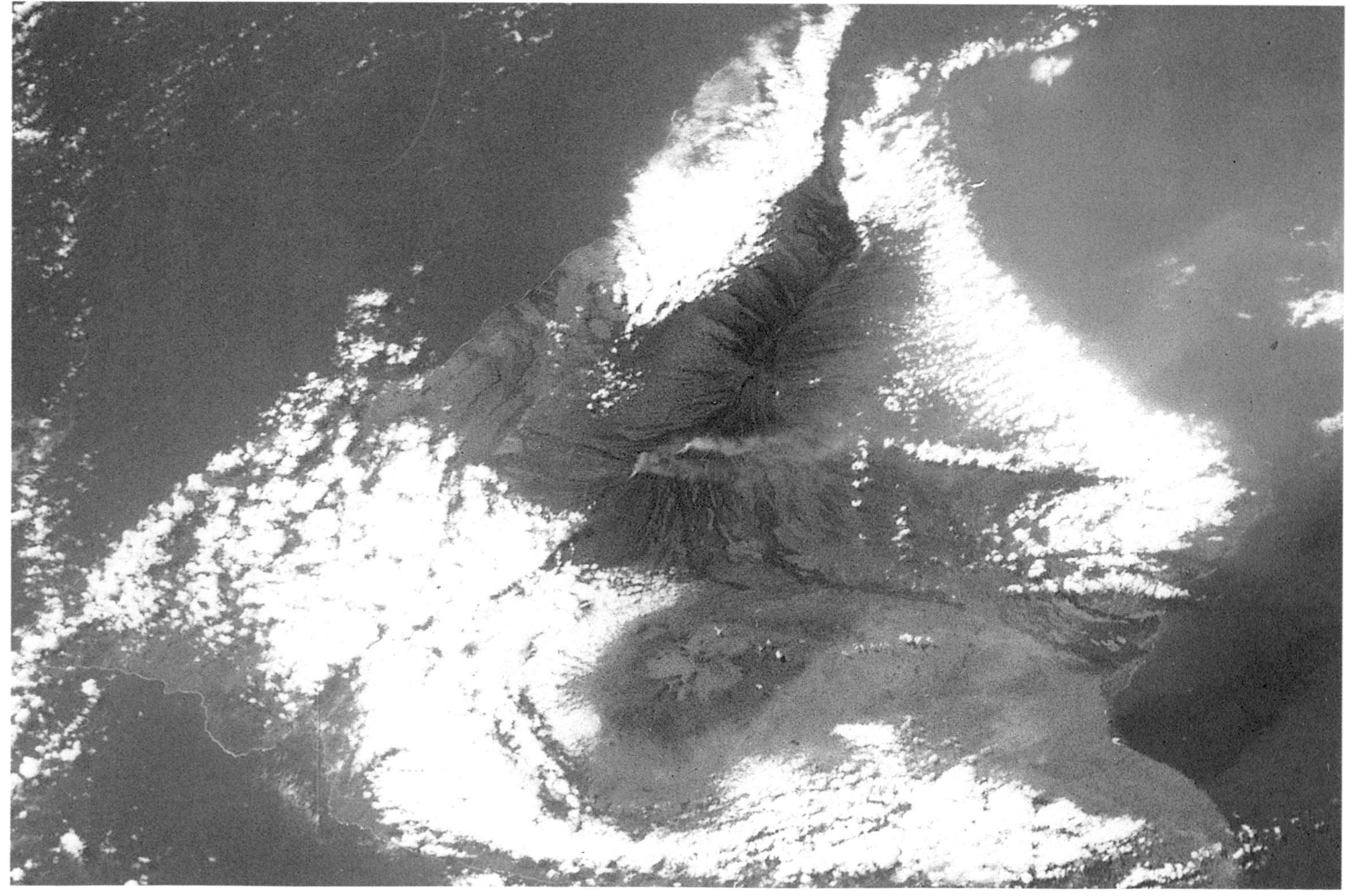

▶ Volcanoes in eruption
Fumes from the lava spewing out of the erupting Mauna Loa volcano on Hawaii. The astronauts in STS 41-C photograph the eruption in April 1984. Mauna Loa erupts frequently, unlike Mauna Kea below, which has not erupted for thousands of years. Hidden beneath the clouds to the left of the picture is another erupting volcano, Kilauea.

on this mission. The other two, of SBS-4 and Telstar 3-C, were spring-launched. After the third had been sent spinning into space, elated mission specialist Richard Mullane exclaimed to Mission Control: 'We're three for three!'

Satellites are also launched – and retrieved – by the RMS robot arm. This is a 15-meter (50-foot) long jointed arm, which extends from inside the payload bay on the port side of the orbiter. It has three flexible joints – a 'shoulder', where it is attached to the orbiter, an 'elbow' and a 'wrist'. At its 'hand' end is a device called an end effector, which has a set of wire snares to grip with. Satellites designed for launch and retrieval with the arm have a docking, or grapple, pin attached.

Payload stations

When the shuttle astronauts are launching satellites, they work at the aft crew station at the rear of the flight deck of the orbiter. They can look forwards into the payload bay through two windows and upwards into space through two more. The pilot (or commander) stands to the left, where he can reach the controls that can rotate the orbiter or move it fore and aft. These controls, the rotational hand controller and the translational hand controller, are duplicates of controls in the cockpit, which the commander and pilot use

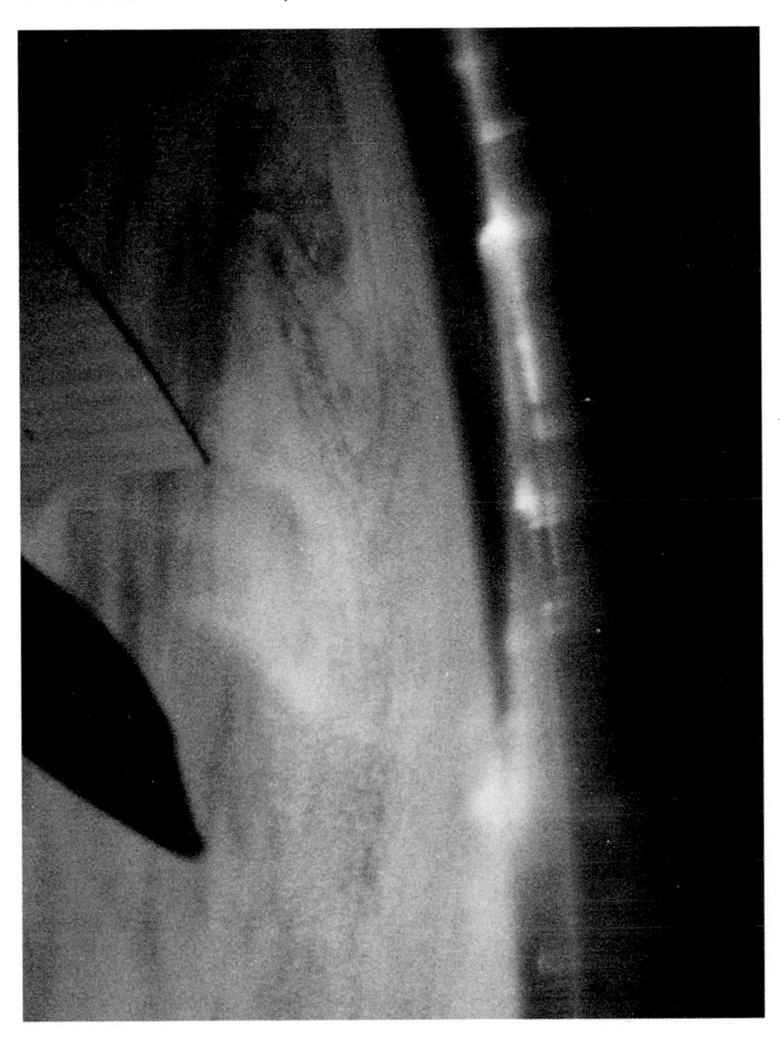

▼ (left) **Northern lights**
A unique perspective of the aurora borealis presents itself to the crew of STS 51-B. The shimmering curtains of colored light result from the interaction of charged particles of the solar wind with the molecules of air in the tenuous upper atmosphere.

▼ **Greening the desert**
STS 41-C astronauts snap this fascinating picture in the desert of Saudi Arabia, near Al Hufuf. The desert is blooming in circular patches where center-pivot sprinklers are employed to irrigate the land.

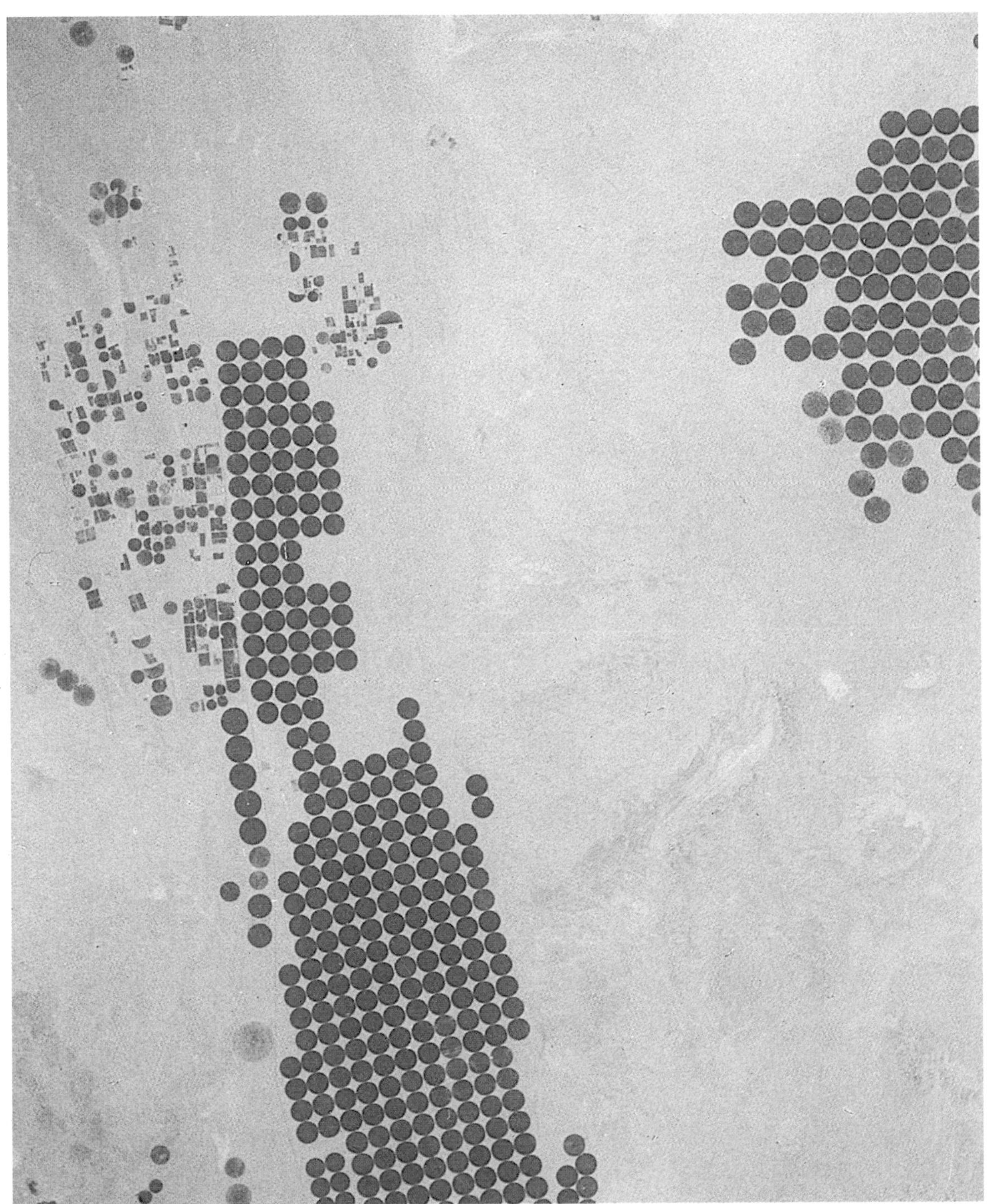

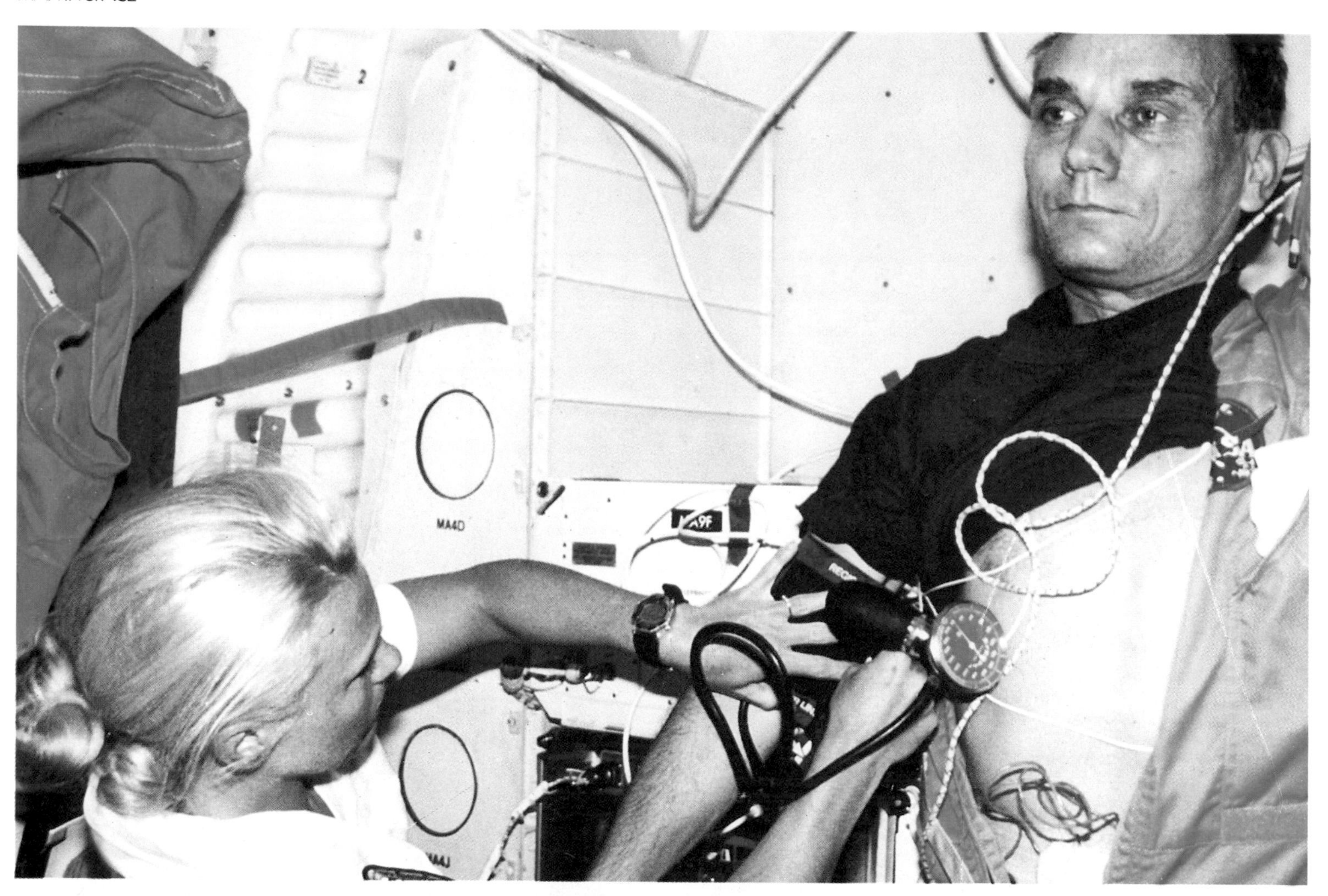

◀ **Noises off**
Another occasional astronaut, Senator Jake Garn, hitches a ride aboard *Discovery* on STS 51-D. For NASA this is perhaps a wise move for Garn is head of the government committee in charge of its budget. Here, he is submitting stoically to the attentions of Rhea Seddon taking his blood pressure. His stomach is 'wired for sound', and he involuntarily obliges by succumbing to space sickness.

▶ **Daunting paperwork**
Dr William Thornton checks the roll of data resulting from a series of medical experiments he has carried out on the crew of STS 8. He is wondering if there is another environmental hazard in orbit – being throttled by data rolls!

▶ (far) **Growing crystals**
In the mid-deck area of *Discovery* on STS-26, George ('Pinky') Nelson is photographing one of the crystal growth experiments.

◀ (below) **Superfast, superpure**
Industrial astronaut Charles Walker tends the continuous flow electrophoresis experiment on *Discovery*'s mid-deck during STS 41-D. In the experiment electric current is applied to help the separation and purification of pharmaceutical chemicals.

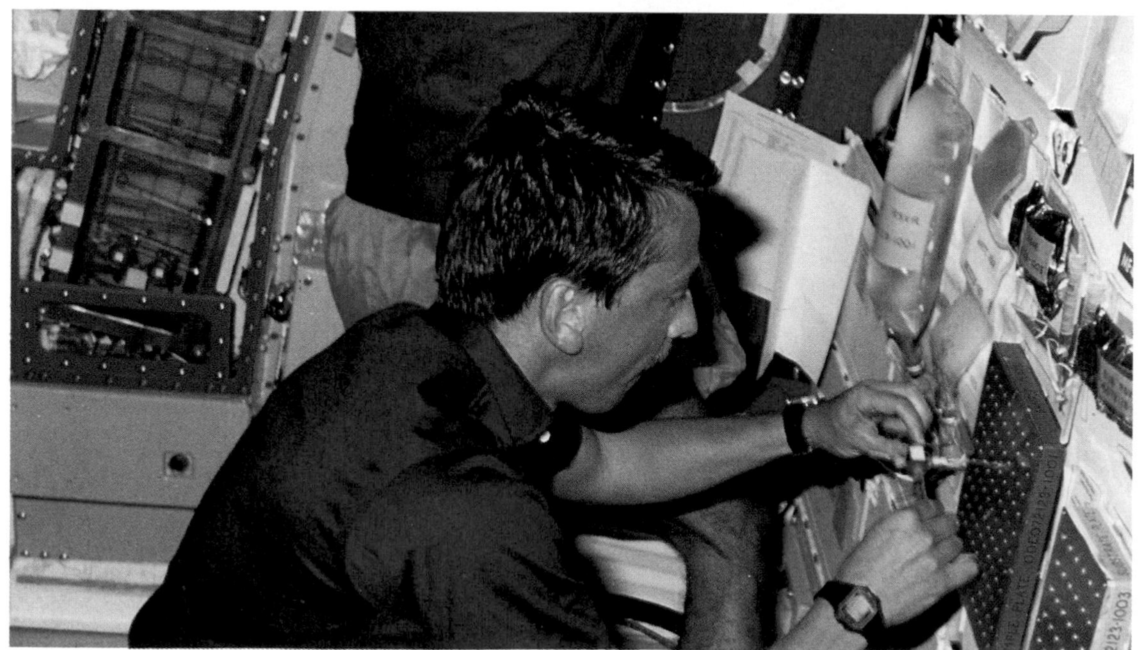

when flying the orbiter normally.

The mission specialists who carry out the launches stand to the right of the pilot. If appropriate, one operates the hand controls that work the RMS arm. To the upper right is a pair of closed-circuit television monitors linked to television cameras inside the payload bay and on the RMS arm.

Getaway specials

The mission specialists have other tasks to carry out besides launching satellites. On many missions they tend, where appropriate, the so-called getaway specials (GAS). These are essentially self-contained experiments that are accommodated in canisters at the sides of the payload bay. NASA charges a few thousand dollars to carry them into space, which gives businesses, research organizations and educational institutions, even schools, economical access to space. A wide variety of experiments has been flown: to make artificial snow crystals (STS-6); to study a live ant colony (STS-7); to examine liquid sloshing behavior (51-G).

Experiments are also conducted inside the orbiter, on the mid-deck. There is not a great deal of room there, but much useful experimentation has taken place, particularly in refining and processing biological and pharmaceutical samples. The apparatus

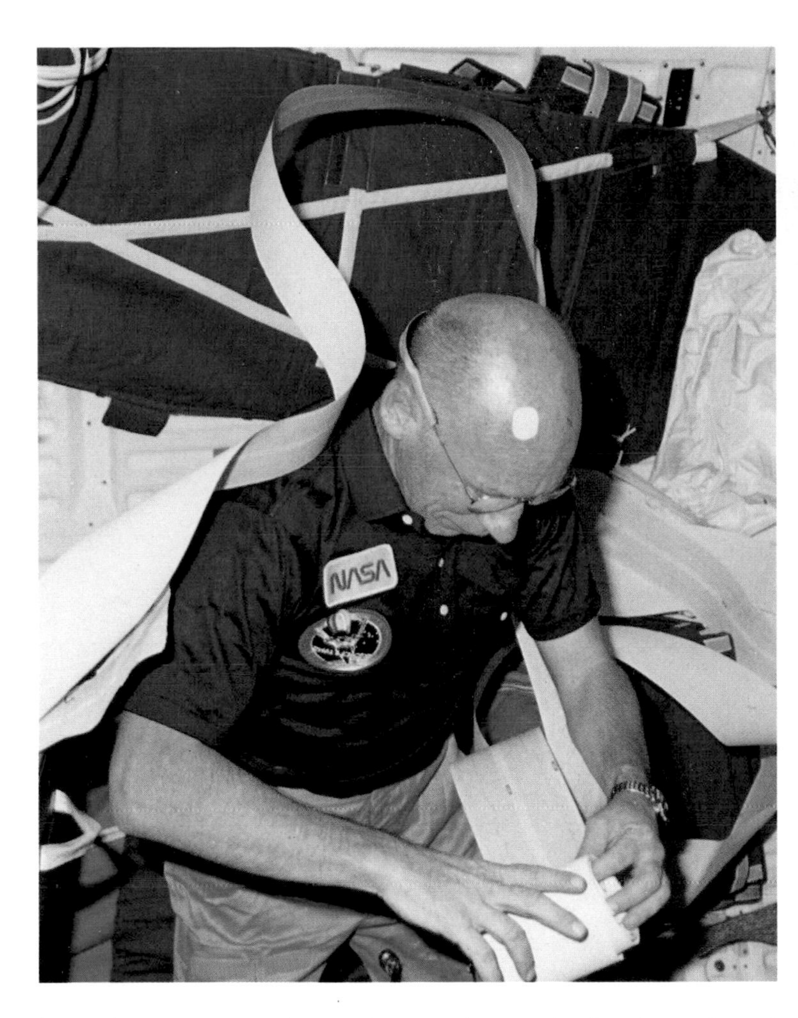

has to be compact and can, of course, be attached to the walls and roof of the cabin as well as the floor!

Crystal gazing

Typical of this kind of investigation was the Protein Crystal Growth (PCG) experiment that flew on STS-26 in September 1988. Mission specialist George Nelson carried out the work, while his colleagues were busy checking over the newly modified *Discovery* on the first post-*Challenger* flight. The purpose of the experiment was to grow crystals of a number of proteins of particular biological importance.

Crystals grown on Earth are distorted by the presence of gravity. In space, crystals are able to grow larger and without flaws since there is no gravity to distort them. Acquiring a perfectly shaped crystal enables researchers to determine better the crystal's structure. Knowledge of the precise structure of a protein's complex molecules holds the key to understanding how it functions and how its function may be altered and controlled.

Crystals were grown of 10 different proteins, including an enzyme called reverse transcriptase. This enzyme is of vital importance to researchers because it seems to hold the chemical key to the replication of the deadly AIDS virus. It is hoped that more precise knowledge of its structure will lead to the development of more effective drugs to treat AIDS. Another protein investigated was alpha interferon, a substance that has had spectacular

success in treating some forms of cancer.

Other mid-deck experiments on shuttle flights center on space medicine – on the effects of weightlessness on the human body. Mostly, however, such studies are pursued in the dedicated science laboratory-in-space, Spacelab, which affords more room and carries a crew whose energies are devoted solely to scientific investigation (see page 194).

Returning home

After the crew have carried out their allotted tasks in orbit, they get ready for their return to Earth. As a preliminary, they first have to clean up their home in space. A crew of five or more astronauts generates a great deal of garbage even in a few days. This, along with any experimental equipment the crew has been operating, must be carefully stowed and secured. Loose objects that pose no threat when floating about in zero-g can become dangerous when they succumb to the g-forces of re-entry.

In the shuttle the first essential re-entry task is the closing of the payload-bay doors. The orbiter could not re-enter safely with them open. The crew then don an anti-g suit. This is in essence a pair of long pants that can be inflated. When inflated, they exert pressure on the lower part of the body, preventing blood from pooling there when the g-forces begin to bite. If such pooling did occur, it could cause the crew to black out, with potentially disastrous results.

Next it is time to maneuver for the de-orbit burn. The pilot fires the reaction control system thrusters to somersault the orbiter until it is traveling tail-first. Then he fires the orbital maneuvering system engines. This brakes the orbiter, allowing gravity to pull it down from orbit. It is one hour to go before landing. The pilot again fires

▲ **Re-entry glow**
On re-entry into the atmosphere the insulating tiles on the orbiter glow red-hot as a result of air friction. The windshields of the orbiter are lit up by the glow. Here 41-G mission commander Robert Crippen is returning home after his record fourth shuttle flight.

▶ **Dusty touchdown**
Leaving desert dust billowing in its wake, *Atlantis* rolls along the runway at the Edwards Air Force Base in California at the end of STS-27 on 6 December 1988. The success of this four-day mission, which saw the launch of a classified spy satellite, augers well for America's future space program.

the thrusters to tilt the orbiter so that it is once more traveling forwards in a nose-up attitude. Half an hour later it slams into the tenuous outer atmosphere at a height of about 122,000 meters (400,000 feet) and at a speed of about 26,500 km/h (16,500 mph).

The drag of the air first gently, and then viciously brakes the craft. Its outer skin heats up, and the crew experience g-forces of 3g or more. For about a quarter of an hour, there is a communications blackout between the crew and Mission Control. When the orbiter emerges from the blackout, it is no longer a spacecraft but an aircraft. Slowing all the time, it performs a series of roll reversals to reduce speed still further before gliding in to a runway landing.

Dicing with death

Re-entry and lift-off vie with each other as the most dangerous period of any space mission, rather as take-off vies with landing as the most dangerous part of an airplane flight. Both these phases of space flight have extracted their cost in terms of human life. And there have been in addition several close calls.

At the time of writing there have been two tragedies during

United States

▲ **The *Challenger* seven**
The seven crew members of STS 51-L, who became America's first in-flight space martyrs. At top, from left to right, are Ellison Onizuka (who had flown previously on 51-C), rookie 'teacher-in-space' Christa McAuliffe, rookie Greg Jarvis, and Judy Resnik (41-D). At bottom, from left to right, are rookie Mike Smith, Dick Scobee (41-C) and Ron McNair (41-B). Said President Reagan during a memorial service for the crew at the Johnson Space Center: 'We bid you goodbye, but we will never forget you.'

re-entry, both among Soviet cosmonauts. In April 1967 Vladimir Komarov was killed while returning from a successful orbital mission in the new Soyuz spacecraft (see page 62). In June 1971 Georgi Dobrovolsky, Vladislav Volkov and Viktor Patseyev were killed in Soyuz 11 while returning from a record-breaking visit to the Salyut 1 space station (see page 188).

The most public space tragedy, however, took place at lift-off, and this time it was American astronauts who paid with their lives the penalty for daring to push back the unforgivingly hostile frontier of space. That tragedy took place on 28 January 1986 in the clear blue skies above Cape Canaveral, which had witnessed, though perhaps surprisingly, no fatalities since the Space Age began.

It was the 10th lift-off of shuttle orbiter *Challenger* on mission 51-L. Aboard was a crew of no less than seven. Seventy-three seconds into the flight, the shuttle stack erupted into an obscene, angry orange, fireball. The two solid rocket boosters emerged from the expanding cloud of flame and smoke, but *Challenger* did not. It had been ripped apart.

In an instant commander Richard Scobee; pilot Michael Smith; mission specialists Judith Resnik, Ellison Onizuka and Ronald McNair; and payload specialists Gregory Jarvis and Christa McAuliffe met their untimely deaths. President Reagan, speaking soon afterwards, said: 'They slipped the surly bonds of Earth to touch the face of God.'

Prelude to disaster

In the fall of 1985 the shuttle fleet of orbiters reached its full strength of four with the successful maiden launch on 30 October of *Atlantis* on a classified mission (51-J) for the Department of Defense. On 30 October *Challenger* carried the largest crew ever, eight, into orbit on an impressive Spacelab flight, Spacelab D-1, funded mainly by West Germany and carrying three European astronauts. Then it was *Atlantis*'s turn again on 27 November. This mission, 61-B, saw some spectacular spacewalking as the

▼ **Tell-tale puff**
With *Challenger* only meters above the launch pad, smoke puffs from the lower joint in the right-hand solid rocket booster. The O-ring is not doing its job.

▶ **Fireball!**
A fireball erupts around the shuttle 73 seconds into the flight, 16 km (10 miles) high. *Challenger* is blown to smithereens. One of the solid rocket boosters, still at maximum thrust, streaks heavenwards. The unthinkable has happened.

astronauts assembled modular structures in the orbiter payload bay. NASA called it 'the best flight ever'. It looked as if the shuttle program at long last was getting into its stride.

But then it all began to go wrong again. The scheduled 18 December launch of *Columbia* was postponed again, and again, and again, either for niggling technical problems or for bad weather. Not until 12 January did this 'mission impossible', 61-C, streak into orbit. The delay with *Columbia* held up the next scheduled mission, 51-L, which was dubbed the 'teacher's flight' because it was to carry teacher Christa McAuliffe into orbit to conduct the first lessons from space.

'Go at throttle up'

Mission 51-L, the space shuttle's 25th flight, was in its turn delayed by a succession of minor mishaps. But on 28 January the launch go-ahead was given despite the fact that the previous night had seen temperatures plummet to −2°C and icicles festoon the

▶ Recovering the wreckage
Two days after the *Challenger* disaster, chunks of wreckage are offloaded at Cape Canaveral Air Force station. This is a section of the underfuselage, which still has some black tiles attached.

▼ Homage
At the astronauts' home base, the Johnson Space Center, Houston, wreaths are laid in memoriam. The President leads the nation's mourning at a memorial service there. 'This America', he says, 'was built by men and women like our seven star voyagers, who answered a call beyond duty.'

launch gantry. With teacher McAuliffe on-board, millions of children at schools throughout the nation followed the final countdown.

At 11.38 am local time *Challenger* roared off the pad, to the accompaniment of cheers at the Cape and in hundreds of classrooms nationwide. It performed its preprogramed roll as it headed for space, and at 52 seconds into the flight Mission Control radioed: '*Challenger*, go at throttle up.' The main engines were now putting out maximum thrust ('max Q'). It seemed like a perfect lift-off. But, unseen by the excited shuttle watchers below, a plume of flame had emerged from the lower joint of the right-hand SRB. Mission Control and the crew were equally oblivious of this dangerous development. Said commander Scobee at 70 seconds, acknowledging Mission Control: 'Go at throttle up.'

Challenger was now traveling at nearly twice the speed of sound. The plume of flame was playing like a blowtorch on the lower strut that held the right-hand SRB to the external fuel tank. At 72 seconds the strut severed. The SRB swung round on the upper strut and smashed into the tank. The leaking fuel ignited in a gigantic explosion. Exclaimed pilot Smith at 73 seconds: 'Uh-oh!' Then silence. *Challenger* was no more. Her crew had become the first in-flight casualties in the American space program after 55 successful missions.

Flawed decision-making

The nation, the world, mourned. President Reagan appointed a 13-member commission, including astronauts Neil Armstrong and Sally Ride, to look into the cause of the disaster and to recommend steps to be taken which would ensure that such a tragedy would never happen again. The chairman was former Secretary of State William Rogers. The Rogers commission conducted painstaking and wide-ranging investigations into the circumstances surrounding the disaster, considering not only the technical minutiae but also the process of decision-making within NASA and the contractors involved in the shuttle program. The commission reported back to the President in June.

The immediate cause of the *Challenger* disaster was pinpointed as the failure of a seal, or O-ring, in the lower joint of the right-hand SRB. This allowed the escape of the hot gases generated by the burning solid propellant. And it was this gas that set in motion the chain reaction that led to the fireball explosion.

It came to light that potential problems with the O-ring seals had been identified years before. And 12 months earlier the O-rings in the 51-C boosters suffered extensive damage on lift-off. That launch had been made following a night when temperatures had fallen way below zero – near-identical conditions to those on the night preceding the *Challenger* launch. The explanation was simple: the rubber O-rings lost their resiliance – their ability to expand into the joints they were designed to seal – at low temperatures.

It was with this in mind that engineers at the SRB manufacturers Thiokol had opposed the launch of *Challenger* on the frosty morning of 28 January. Engineers at the orbiter manufacturers Rockwell had opposed the launch too because of the potential danger of ice from the gantry falling on to and damaging the heat-shield tiles. Yet their advice had not been communicated to the launch team. So *Challenger* had lifted off and had been blasted out of the sky.

The Rogers commission report was scathing in its criticism of NASA's management structure and the lines of communication which led to the flawed decision-making that allowed 51-L to launch. Among its recommendations were a new seal design; a crew escape system; modification of orbiter landing gear, brakes and steering system; and the overhaul of mission management structure. Newly appointed NASA Administrator James Fletcher, recalled for a second term in office, was charged with the task of implementing the report's recommendations. This he did over a 22-month period. And in September 1988 the shuttle program resumed with the flight of *Discovery* on STS-26 (see page 95).

The *Challenger* disaster underlined what every astronaut and cosmonaut knows only too well. That space flight is, and will always be, a deadly business. There really can be no such thing as a routine safe flight. Astronauts ascend to the heavens on a pillar of fire, sitting on thousands of tonnes of high-explosive fuel. In orbit they are a few centimeters away from a radiation-saturated, meteoroid-filled vacuum. On their descent to Earth, they run the fiery gauntlet of re-entry and then trust that the heat shield will do its job and on Soyuz that the retrorockets will fire properly and the landing parachutes will open.

First American in space John Glenn spoke of the fears of the early astronaut corps when he said after the *Challenger* accident: 'We always knew there would be a day like this.' And without doubt there will be more such days in the future as man attempts to push back further the space frontiers. But, as President Reagan so eloquently put it: 'Sometimes when we reach for the stars, we fall short. But we must press on despite the pain.'

▼ The last goodbye
The remains of the 'seven star voyagers' are air-lifted from the Kennedy Space Center on 29 April 1986. The nation still asks, why did they have to die?

Chapter 5

WALKING IN SPACE

◀ **Human satellite**
Jetting hundreds of kilometers above his home planet, Bruce McCandless on 7 February 1984 becomes the first human satellite. He is test flying the latest piece of EVA hardware, the MMU.

▶ **Skylab EVA**
During the first Skylab occupancy, beginning 25 May 1973, Owen Garriott makes a routine spacewalk to deploy an experiment outside the 90-tonne space station, which was almost crippled at launch.

Astronauts face danger every time they venture into space, even when they are cocooned inside a spacecraft precisioned-engineered for the purpose. But sometimes missions call upon them to leave their craft and work outside in the unforgiving vacuum of space. Such out-of-vehicle work is properly termed extravehicular activity, or EVA, but is popularly known as spacewalking.

Alexei Leonov for the Soviets and Edward White for the Americans tentatively pioneered this ultimate and most hazardous space experience in 1965. On many occasions since then spacewalking astronauts have demonstrated their versatility in effecting repairs to damaged and ailing spacecraft. They saved the Skylab missions from disaster and extended the lifespan of the Salyut space stations.

Most spectacularly, whizzing about in Buck Rogers style jet-propelled backpacks, they have ushered in an era of in-orbit satellite servicing. They are able to retrieve, mend or recover satellites that have malfunctioned. In just a few years time they will assume the role of construction engineers when the international space station *Freedom* gets off the ground.

This chapter traces the genesis of spacewalking in orbit from Voshkod through Salyut. The rather different low-gravity spacewalking involved in the Apollo Moon-landing program is considered in the next chapter.

◀ **'A million dollars'**
When Edward White emerges from Gemini 4 to make the first US spacewalk on 3 June 1965, he says the view 'must be worth a million dollars'. This fine shot, showing White with camera and jet gun, is one of the photo greats of the Space Age.

▼ **The first spacewalker**
Alexei Leonov floats outside his Voshkod 2 capsule in raw space on 18 March 1965 to pioneer the hazardous practice of spacewalking.

ALEXEI LEONOV WAS SWIMMING, swimming in space. No one had experienced such a sensation before. No astronaut or cosmonaut before had quit his spacecraft to meet the universe face to face. He was the first person to walk in space, to perform extravehicular activity. The date was 18 March 1965; the spacecraft was Voshkod 2.

Voshkod was about to begin its second orbit of the Earth when Leonov got ready to boldly go where no man had gone before. His colleague in Voshkod was Pavel Belyayev, who helped Leonov prepare for his encounter with the unknown. Leonov exited Voshkod in two stages through an inflatable telescopic airlock which extended into space. Having checked out his four-layer spacesuit in the airlock, Leonov reported to Belyayev: 'This is Diamond. Everything is in order.'

Then the exit hatch was opened and Leonov floated out into a world of blinding sunlight and velvety blackness, linked to his spacecraft only by a 5-meter (16-foot) long tether, or umbilical, which also fed him oxygen to breathe. Radioed an excited Belyayev to ground control: 'A man has stepped out into cosmic space.'

Voshkod was traveling over the Black Sea at the time. Leonov reported that he could see the Caucasus mountains. For just over 10 minutes Leonov swam, somersaulted and gyrated at the end of the umbilical. Later he said: 'I felt absolutely free, soaring like a bird as though I was flying by my own efforts.... The view of the cosmic expanse so enthralled me that there was no place in my mind for other feelings.'

After his triumphant time outside, Leonov almost came to grief when trying to get back into the airlock hatch. First the movie camera that had recorded his epic walk got stuck in the opening. Struggling to free it, he began sweating profusely as he overexerted himself. His suit systems could not cope, and the sweat got into his eyes and fogged his helmet. Then when he managed to free the camera, he found his suit had ballooned out so much that he could not get back through the hatch. Not until he had reduced the oxygen pressure inside his suit to a dangerously low level (a quarter of an atmosphere) was he able to squeeze back inside.

The saddest moment

Leonov's 10-minute walk proved that man could, at least for a short time, survive outside in space, despite the high vacuum, the intense heat of the Sun and the freezing cold of shadow. But there appeared to be a potential problem with overexertion, a problem that would soon recur.

Spacewalking was an essential objective of the American Gemini program, which got underway with the launch of Gemini 3 just five days after Leonov's triumph (see page 51). America's first spacewalking attempt took place on the second Gemini mission, Gemini 4, on 3 June 1965. On the third orbit of the Earth Edward White opened the hatch of the depressurized spacecraft and floated into space. Like Leonov, he was connected to Gemini's life-support system by an umbilical/tether.

White carried a jet gun (HHMU, hand-held maneuvering unit) to help him maneuver when floating free. But after four minutes it ran out of fuel, and thereafter the only way he could move about was by pulling on the umbilical. As he tumbled with effortless ease, he gave a running commentary: 'This is the greatest experience; it's just tremendous.... Right now I'm standing on my head and I'm looking right down... looks like we're coming up on the coast of California.'

All too soon for White, the extravehicular spectacular had to end. As he made his way back to the hatch he said: 'This is the saddest moment of my life.' Back inside, he had difficulty in closing the hatch. He eventually achieved this with fellow crew member James McDivitt holding on to his legs. The exertion caused White's pulse to race to nearly 180 beats a minute. Sweat poured from his face and his vizor fogged up. Leonov's problem.

Spacewalking around the world

After White's walk in space, priority was given over the next four Gemini missions to achieving other objectives, such as rendez-vousing with spacecraft in orbit and docking with Agena target vehicles. The next spacewalk was planned for Gemini 9, which

◀ **At the busy box**
Gemini 12 spacewalker Edwin Aldrin at last cracks the EVA problem by carefully pacing himself through 19 assigned tasks on 12 November 1965. Here he is seen working at a busy box on the Agena target Gemini is docked with.

▶ **Deep spacewalk**
On 17 December 1972 Apollo 17 pilot Ronald Evans exits the CSM *America* for an hour to retrieve film cassettes from external cameras. *America* is still 290,000 km (181,000 miles) from home. It is only the second deep spacewalk ever made. Thomas Mattingly had made the first deep-space EVA from the Apollo 16 CSM *Casper* the previous April.

proved to be a jinxed mission from the outset. The prime crew for the mission, Elliott See and Charles Bassett, were killed when their T-38 jet trainer crashed on the last day of February 1966. The back-up crew, Thomas Stafford and Eugene Cernan, eventually flew into space on 3 June. The jinx descended again when they were unable to dock with their Agena target, which looked like 'an angry alligator' because a shroud did not jettison from around the docking port.

On the second day of the three-day mission, Cernan exited through the hatch for an ambitious spacewalk scheduled for 167 minutes, nearly two orbits of the Earth. But the same problem that plagued the two previous spacewalkers returned. Cernan found it very difficult to control his movements. The slightest tug against the umbilical or push against the spacecraft set him tumbling.

The sweating began, so did the fogging of the vizor. When Cernan attempted to strap on a Buck Rogers type jet pack (the AMU, astronaut maneuvering unit), things went from bad to worse. His pulse raced and he was almost blinded by sweat and the fogging. He was ordered back inside early. Nevertheless he set a new EVA record of 2 hours 8 minutes.

The snake house

The next mission, Gemini 10 (18-21 July 1966), saw Michael Collins spacewalking twice but with less ambitious objectives. His companion was John Young, making his second Gemini flight. On the first occasion on day two of the mission, he conducted a simple 'stand-up' EVA, emerging up to the waist through the open hatch. The main object was to take photographs. He later vividly described what it was like. Looking out of the hatch as Gemini traveled through the Earth's night-time shadow, he said: 'The stars are everywhere.... The planet Venus appears absurdly bright.... Down below the only light comes from lightning flashes.'

After about 50 minutes Collins's eyes began to water, as did Young's, and the EVA was terminated. The eye watering proved to be caused by a leakage of lithium hydroxide from the canister that absorbed carbon dioxide from the suit life-support system. On day three of the mission Collins quit the hatch on an EVA to retrieve an experiment package from an Agena vehicle that had been in space for four months. Young maneuvered Gemini into formation with the Agena, and Collins drifted across to it at the end of a 15-meter (50-foot) long tether, using a jet gun to maneuver.

All went reasonably well until he started back to Gemini, pulling on the tether. Gradually he became more and more entangled. And he got into a real mess when he finally got back into the hatch. As he struggled to untangle himself, he said: 'This place makes the snake house at the zoo look like a Sunday school picnic!'

Pooped again

Collins was whacked by the time he had sorted himself out. Spacewalking then was proving to be a particularly difficult aspect of life in space. How was this difficulty to be resolved, that was the question?

The penultimate Gemini mission, Gemini 11 (12-15 September 1966), provided no clue to the problem either. The mission required the crew to dock with an Agena target and then for Richard Gordon to spacewalk over to the Agena and attach a tether to it. The two spacecraft would then be separated until the tether was taut, and then the two craft would be set spinning in an attempt to create artificial gravity.

A 155-minute spacewalk was planned, but the deceptively simple matter of attaching the tether made Gordon hot and sweaty and fogged up his vizor. It was the same old story. His colleague Charles ('Pete') Conrad encouraged Gordon, perched astride the Agena, with a cry of 'Ride 'em cowboy!', but called him back as soon as he had tied the tether. After just 38 minutes EVA, Gordon confessed: 'I'm pooped Pete.'

One more Gemini flight to go, and NASA just had to beat the EVA barrier. So when Edwin Aldrin stepped out into space from Gemini

▶ **In the water tank**
Both American and Soviet spacewalkers rehearse for their EVAs in water tanks big enough to hold spacecraft mock-ups. In weighted suits they achieve a neutral buoyancy state approximating to that of weightlessness. Here a Soviet cosmonaut practices in the hydrotank at the Yuri Gagarin Cosmonaut Training Center in Star City, near Moscow.

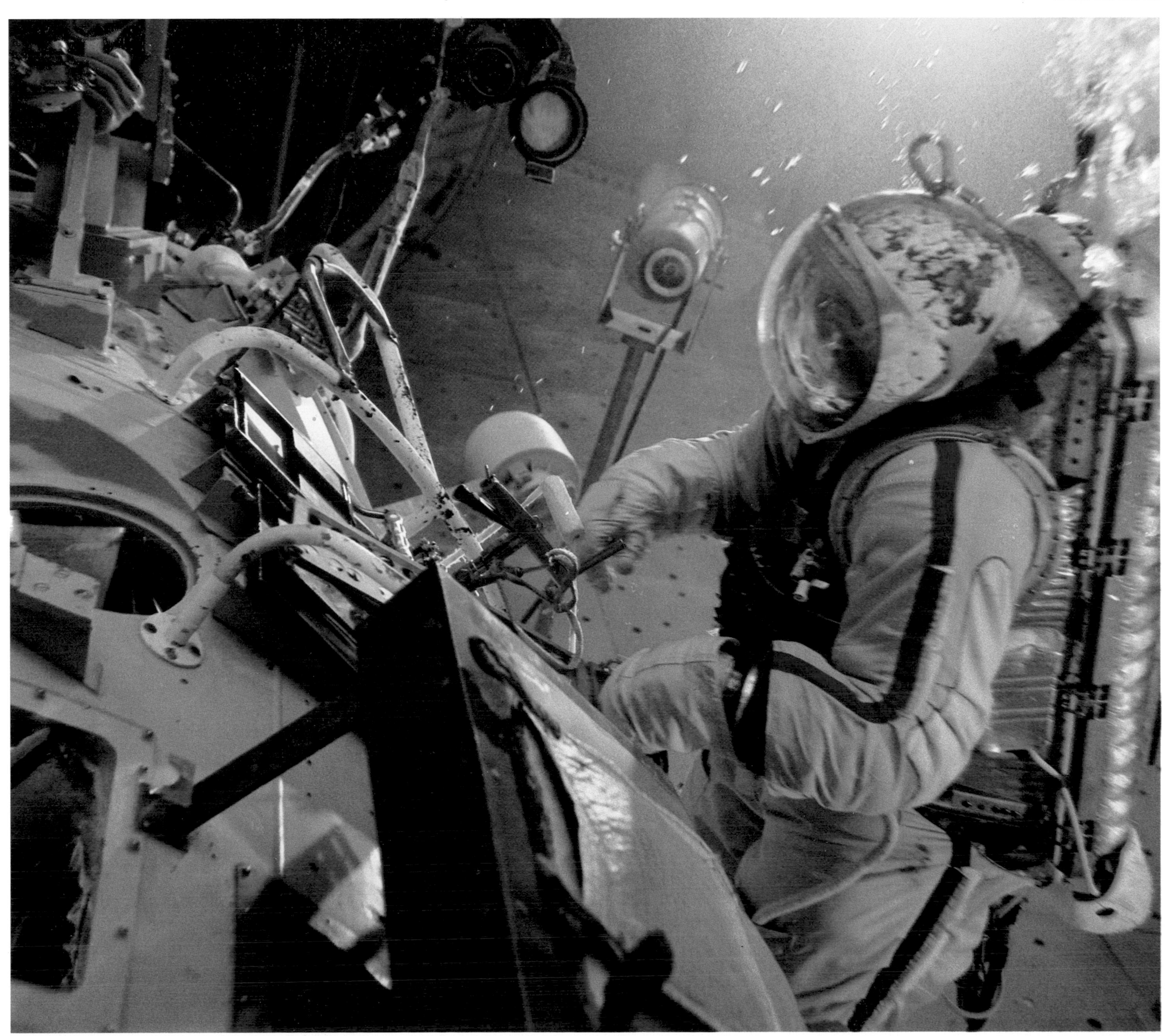

▶ One Story high
The new shuttle spacesuit gets its first airing in April 1983 on STS-6. Story Musgrave is shown here safely tethered, floating above the payload bay. With a reference to his water-tank training, he says: 'The tank's a little deeper than the one I'm used to!'

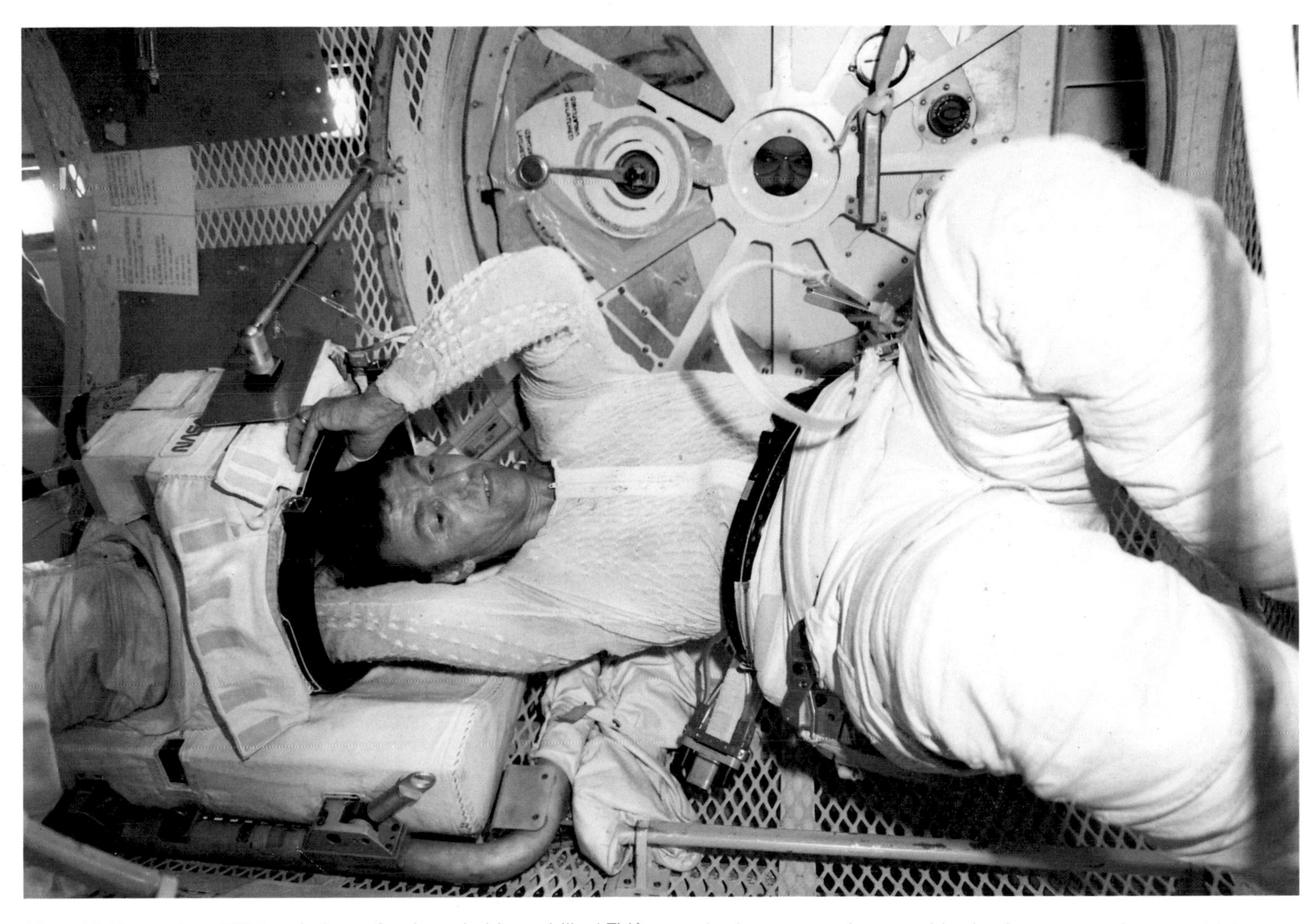

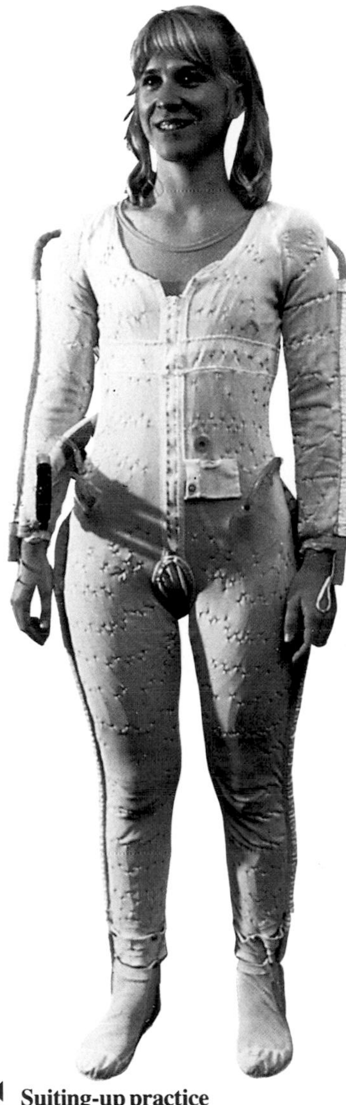

◀ **Suiting-up practice**
Joe Engle practices donning the space shuttle spacesuit in the fleeting weightlessness of an arcing airplane flight. He puts on the trousers first and then floats up into the upper torso. Note the rigid waistband of the trousers, which connects and seals with a similar band on the torso.

▲ **Liquid-cooling garment**
Underneath the spacesuit the astronaut wears these water-cooled long johns. The pipes channel water into and out of the garment from a supply in the portable life-support system, or backpack, which is built into the upper torso of the spacesuit.

12 on 13 November 1966, two days after launch, his umbilical EVA was planned virtually to the very minute. It adopted a softly, softly approach, with regular two-minute rest periods. This enabled Aldrin to get through his workload with no hassle, no racing pulse, no sweat, no fogging in an EVA lasting over two hours. Wherever he worked, he anchored himself using a portable handhold and waist harness, attaching them by means of Velcro patches to mating patches stuck all over the spacecraft. At a 'busy box' at the rear of Gemini he used tools to turn bolts and perform other tasks.

Along with two stand-up EVAs, one before and one after the umbilical EVA, Aldrin on this mission spent over 5½ hours spacewalking. The next time he stepped out into space it would be spacewalking of a very different kind, and it would take place on the Moon (see page 164).

The spacesuit evolves

The original spacesuit worn by the Mercury astronauts was not really a spacesuit at all. It was a silvery version of the high-altitude pressure suit worn by military pilots. (The silvery finish was by all accounts chosen to make the astronauts look more like the popular picture of spacemen!) The Mercury spacesuit was made up of two main layers, an inner rubberized pressure layer, which was supplied with oxygen under pressure; and an outer restraint layer to prevent the inner layer ballooning.

The Gemini astronauts were true spacemen in that they ventured bodily into space on EVA. They wore a suit with extra layers – four in all – to give them protection against the hazardous space environment. When on EVA, their suits were connected by an umbilical to the spacecraft life-support system.

The Apollo spacesuit was designed for use either with an umbilical when the EVA took place in space, or with a portable life-support system (PLSS) backpack for lunar surface EVA. The basic suit had a three-layer construction – an inner comfort layer, a pressure 'bladder' and a restraint cover to prevent ballooning. The joints of the suit were made flexible at the body joints. And a built-in system of cables and a kind of block-and-tackle pulley system was incorporated, which allowed the astronauts to move their limbs more easily.

For most of the time in space the astronauts wore the basic three-layer suit over a constant-wear garment, which was like a pair of 'long-johns'. On top there was a protective three-layer outer suit. For the lunar surface EVAs the astronauts swopped the

◀ **Reflections**
The orb of the Sun and the nose of *Challenger* are reflected in the vizor of Bruce McCandless on his pioneering test-flight of the MMU on mission 41-B. He takes it easy, saying: 'I could go faster, but why rush it?'

▼ **Buck and Flash**
Robert Stewart flies the MMU this time, with obvious enjoyment. Shuttle commander Vance Brand says of *Challenger*'s two human satellites: 'They call each other Flash Gordon and Buck Rogers.'

constant-wear garment for a liquid-cooling garment. This had a network of plastic pipes interlaced in it, through which cooling water was circulated from a supply in the PLSS backpack. This was necessary to prevent the astronaut from overheating when working. Also for the surface EVA, the suit had a thicker protective outer suit, made up of 17 layers of plastic film, neoprene-coated nylon and Beta cloth, a fireproof fiberglass fabric.

Altogether, the whole EVA suit plus backpack was called the extravehicular mobility unit (EMU). It provided life-support for a four-hour EVA. In all it weighed on Earth 86 kg (190 pounds), a tough load to carry for hours at a time. Fortunately the lunarnauts had to shoulder only one-sixth of this because of the low lunar gravity.

The shuttle EMU

Whereas the early astronauts wore spacesuits for most of the time in space, the shuttle astronauts wear ordinary clothes in a shirt-sleeve environment much like that in an airliner. Only when they go on EVA do they don a spacesuit. The shuttle EMU is developed from the Apollo version. It has similar multilayer construction, but with significant differences. It comes in two parts – upper torso and trousers, or lower torso. The upper torso has a rigid frame of aluminum, and the PLSS backpack is permanently attached to it. The trousers are flexible and join at the waist with the upper torso by means of a connecting ring and seal.

On the shuttle two spacesuits are usually carried and stored in the airlock on the mid-deck. This is a cylindrical chamber with two hatches, one leading from the crew cabin, the other leading into the unpressurized payload bay.

The suiting-up operation in the airlock is relatively simple and takes only about 10 minutes. But the EVA astronauts have to spend at least two hours in the airlock beforehand breathing pure oxygen. This is necessary to flush nitrogen out of the blood. If nitrogen were still in the blood when they switched to the

▲ **The flying armchair**
Another view of McCandless's first untethered spacewalk. The MMU has its origins in the AMU (astronaut maneuvering unit), hardware that was designed for Gemini missions but was never tested in space. A later version was tested in space, but within the confines of Skylab in 1973. From this the present MMU design developed.

◀ **On the cherry picker**
Bruce McCandless, after his triumphant space debut on the MMU, now hitches a ride on *Challenger*'s manipulator arm. The contraption attached to it is the mobile foot restraint, or cherry picker. The robot manipulator arm, the shuttle's crane, is proving to be one of the most successful pieces of shuttle hardware.

reduced-pressure oxygen in their suit, it would bubble out and give them a painful and debilitating attack of the bends.

The first test of the shuttle EMU was scheduled for the shuttle's fifth mission (STS-5, 11-16 November 1982). Mission specialists Joseph Allen and William Lenoir were to carry out trial EVAs. After a one-day postponement of the trials because of space sickness, the two astronauts suited up. But they had to scrub the trials when their PLSS backpacks malfunctioned. The 19,000 rpm fan in Allen's suit 'sounded like a motorboat'. Sloppy assembly was later found to have caused the problems.

It was left to Story Musgrave and Donald Peterson on the next mission (STS-6, 4-9 April 1983), *Challenger*'s first, to put the shuttle EMU through its paces. In 3½ hours of EVA the EMUs put in a flawless performance as the astronauts practiced typical operations with tools at designated work stations. All the while they remained tethered to the payload bay.

The flying armchair

The next stage in the development of spacewalking techniques came with the flight of *Challenger* on mission 41-B (3-11 February 1984). This mission was a bittersweet one. On the negative side two satellites, Westar VI and Palapa B-2, costing some $150 million, were lost into useless low orbits. This was balanced on the positive side by show-stopping spacewalks, the like of which had never been seen before.

On 7 February Bruce McCandless and Robert Stewart left *Challenger's* airlock to test-fly a jet-propelled backpack that has been likened to 'a clumsy overstuffed armchair without a seat'. This flying chair is properly called a manned maneuvering unit (MMU). It carries a supply of compressed nitrogen gas, which feeds 24 thruster units located on every side. The MMU is flown by the astronauts squirting gas from the thrusters. Hand controls on the arms fire different combinations of thrusters for movement in different directions. The controls follow the layout familiar to spacecraft crews. The left-hand controller governs fore-aft, right-left, up-down movements; the right-hand controller handles roll, pitch and yaw motions.

It was McCandless's turn first. As he flew the MMU out of the payload bay, he became the first person to spacewalk without a lifeline; the first human satellite. As he 'gunned' the MMU farther and farther away, he said: 'Hey this is neat!... That may have been one small step for Neil, but it's a heck of a big step for me!'

He jetted out to about 45 meters (150 feet) away, and then returned before moving away again to double that distance, when he could be seen as only a bright speck in the blackness. When he came back, McCandless floated in front of *Challenger* and asked

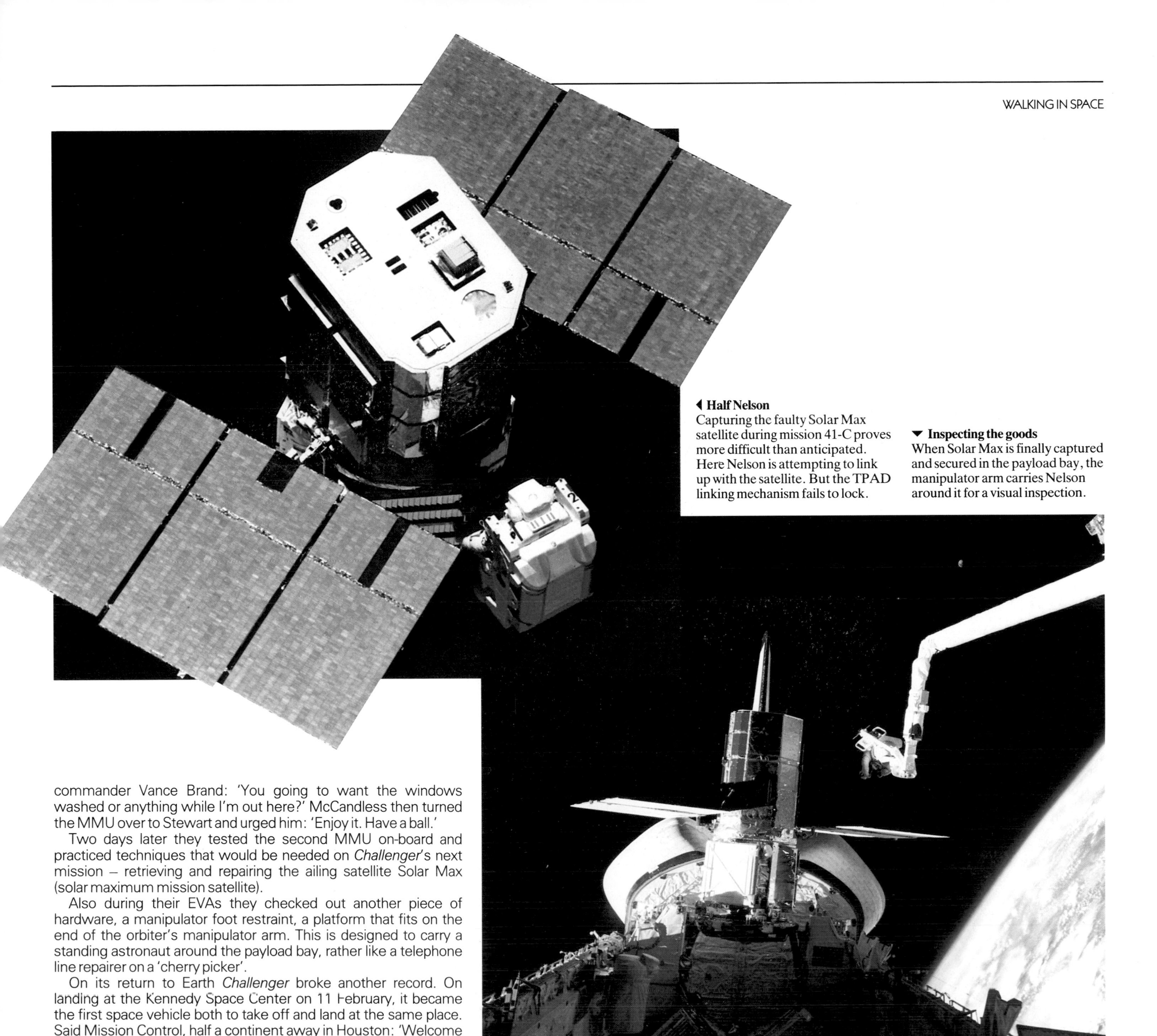

◂ Half Nelson
Capturing the faulty Solar Max satellite during mission 41-C proves more difficult than anticipated. Here Nelson is attempting to link up with the satellite. But the TPAD linking mechanism fails to lock.

▾ Inspecting the goods
When Solar Max is finally captured and secured in the payload bay, the manipulator arm carries Nelson around it for a visual inspection.

commander Vance Brand: 'You going to want the windows washed or anything while I'm out here?' McCandless then turned the MMU over to Stewart and urged him: 'Enjoy it. Have a ball.'

Two days later they tested the second MMU on-board and practiced techniques that would be needed on *Challenger*'s next mission – retrieving and repairing the ailing satellite Solar Max (solar maximum mission satellite).

Also during their EVAs they checked out another piece of hardware, a manipulator foot restraint, a platform that fits on the end of the orbiter's manipulator arm. This is designed to carry a standing astronaut around the payload bay, rather like a telephone line repairer on a 'cherry picker'.

On its return to Earth *Challenger* broke another record. On landing at the Kennedy Space Center on 11 February, it became the first space vehicle both to take off and land at the same place. Said Mission Control, half a continent away in Houston: 'Welcome home. That was a fantastic job.'

Capturing Solar Max

Challenger next soared into orbit on 6 April on the most challenging shuttle mission yet. The main objective of the mission (41-C) was to effect running repairs on Solar Max. This was a Sun-monitoring satellite that had broken down only a few months after its launch in February 1980.

The mission began well with the deployment of a bus-size satellite known as the LDEF (long-duration exposure facility), which carried over 50 experiments devised by 200 scientists in nine countries. The attempt to rescue Solar Max began on day three of the mission. Commander Robert Crippen maneuvered *Challenger* to within 60 meters (200 feet) of the satellite and then George Nelson jetted out to it in the MMU. The idea was for him to dock with its projecting trunnion pin with a so-called TPAD (trunnion-pin attachment device) and stop it spinning. Then it could be grabbed by the manipulator arm and dropped into the payload bay for repair.

But try as he might Nelson could not get the TPAD to lock on to the pin. He tried to stop the satellite spinning with his hands, but this only made matters worse. Nelson returned to the orbiter, which then moved in. Mission specialist Terry Hart then tried to grab the tumbling Solar Max with the manipulator arm, but in vain. It became clear that it could be captured only if the tumbling was stopped.

Ground controllers at the Goddard Space Center worked feverishly to bring about stabilization and after 36 hours finally succeeded. *Challenger* moved in again. This time Hart captured the satellite first time with the manipulator arm and placed it in a cradle in the payload bay. Next day James van Hoften and George Nelson

▶ (above) **Stinger in the tail**
Flying the MMU, Dale Gardner fixes the stinger in the kick motor of the Westar VI satellite on the 51-A mission. Next he will jet it back within reach of Joseph Allen, standing on *Discovery*'s robot arm.

▶ **Handover**
Allen holds the antenna end of the Westar satellite, while Gardner works to detach the stinger. Manhandling a satellite weighing half a tonne proves a doddle up in orbit.

▶ (opposite) **In the bay**
Allen hangs on to the Palapa B-2 satellite before it is stowed into the payload bay. Note on the right the stingers used to effect capture of the satellites.

FRAME 1

FOR
SALE

◀ **Any offers?**
Having successfully recovered Palapa B-2 and Westar VI from their useless orbits, 51-A astronaut Dale Gardner holds up a For Sale sign. His fellow-worker and photographer Joseph Allen is reflected in his vizor.

▶ **Secondhand sats**
Back at the Kennedy Space Center, Palapa and Westar are waiting to be removed from *Discovery*'s payload bay. Later they will be refurbished and offered for sale.

donned spacesuits and went out to repair Solar Max. It took them about 45 minutes to change the faulty attitude-control box.

The next chore, replacing an electronics box, was much trickier for it required the removal of 28 tiny screws. Despite their bulky spacesuit gloves, screw removal was deftly accomplished. At one point Nelson radioed: 'I lost two screws. One went over the side.' But it didn't matter because the new unit was clipped in place. The job done, Solar Max was relaunched and *Challenger* backed away.

During a subsequent TV interview the crew sported T-shirts that labelled them the Ace Satellite Repair Co. Before they returned home they heard the good news: Solar Max was operating normally again. Said ground control repair chief Frank Cepollina: 'The era of the throwaway satellite is over.'

This *Challenger* mission also helped set another record. Its five-member crew plus the six cosmonauts in the Salyut 7/Soyuz T-11 complex also in orbit made up the largest number of astronauts that had been in space together at one time. One of the cosmonauts, Indian 'guest' Rakesh Sharma, became the first person to practice yoga in space. The idea was to investigate whether weightless yoga could help cure space sickness.

Riding the stinger

Just seven months after *Challenger*'s triumph with Solar Max, it was *Discovery*'s turn to feature in a satellite spectacular. The mission, 51-A, set off on 8 November 1984. It carried two satellites in the payload bay and these had to be launched before the other major objective could be carried out – the recovery of the two satellites that had been lost into useless low orbits on the 41-B mission in February, Westar VI and Palapa B-2.

The operation to salvage the two satellites began on 15 November when *Discovery* rendezvoused with the Palapa. The dramatis personae for the operation were spacewalkers Joseph Allen and Dale Gardner flying MMUs, with Anna Fisher inside the orbiter working the manipulator arm.

Also, an ingenious piece of new hardware was involved, which NASA called an apogee kick motor capture device. But because of its shape it soon became known as the stinger. It was Dale Gardner

◀ **Advertising space**
A jubilant 51-A crew poses after the recovery operation: '2 up' refers to the two satellites they carried up into space and launched; '2 down' to the two they have recovered, which they will carry back to Earth. From left to right, they are: David Walker, Dale Gardner, Anna Fisher, Frederick Hauck and Joseph Allen.

who thought up the stinger, the object of which was to provide a grappling pin which the manipulator arm could snare. The stinger was a little over 1.5 meters (5 feet) long and consisted of a spear-like probe designed to fit into the kick motor of the satellites. When inside, an umbrella-like device would expand and provide a grip. The grappling pin was mounted on the other end.

Allen was called upon to use the stinger first. Flying in the MMU, he jetted over to Palapa and fixed the stinger in place. Then, firing the MMU's thrusters, he stopped the satellite rotating and turned it round so that Fisher could reach the grappling pin. 'Come on in, Anna,' said Allen, 'you've got plenty of room.' Ever so slowly the manipulator arm inched in and finally clasped its target.

Gardner snipped off a protruding part of the satellite's antenna so that it would fit into the payload bay. Next he tried to fit a frame with a grapple pin to the antenna end so that the stinger could be removed. But the frame would not fit. So Allen had to hang on to Palapa until Gardner had removed the stinger and secured an adapter ring at the motor end so that it could be clamped into the payload bay for the ride home. With feet anchored on the edge of the payload bay, Allen held the satellite aloft for a full 90 minutes – one orbit of the Earth – before Gardner had finished. Then they together manhandled it into the bay and locked it into position. The whole operation took six hours.

Next day *Discovery* chased its next rogue satellite, the Westar. And soon it was time for the two satellite recovery experts to get back to work. Gardner flew the stinger this time, locked on to the satellite and maneuvered it towards Allen, who was standing on the cherry picker at the end of the manipulator arm. While Allen held the Westar, Gardner removed the stinger and stowed it back in the payload bay. Fisher in the meantime hauled in Allen and his burden. Then the same procedure was followed as with the Palapa, with Allen holding on to the satellite while Gardner fixed the adapter. Soon Westar joined Palapa in the bay.

President Reagan telephoned his congratulations to the crew: 'You demonstrated that we can work in space in ways we never imagined were possible.' Lloyds underwriters in London, who had paid the bulk of the insurance money on the errant satellites, were so delighted that they rang the historic Lutine Bell twice, the traditional signal of a successful salvage operation. They awarded Allen and Gardner their meritorious Silver Medal.

Fly swatting in orbit

Spacewalking was not planned for *Discovery*'s mission 51-D, which began on 12 April 1985. But an unscheduled spacewalk was hastily organized when one of the satellites deployed refused to function afterwards. It was the Hughes-built comsat Leasat 3. This glitch was particularly embarrassing for NASA because on-board was Senator Jake Garn, who had a not inconsiderable influence on the NASA budget!

Leasat 3 flipped out of the payload bay, frisbee style, as planned,

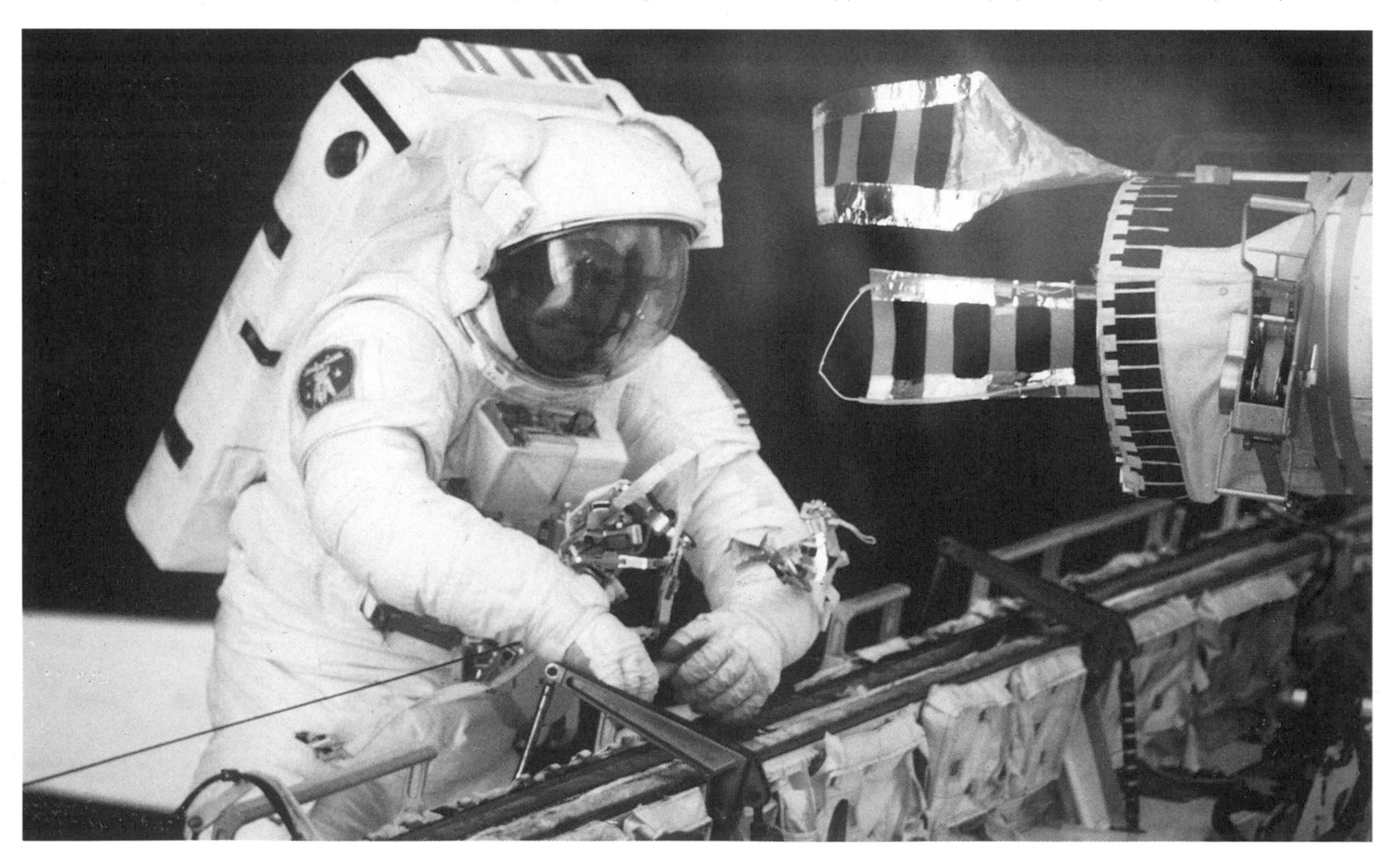

▶ **The flyswatter**
The 51-D astronauts rig up a makeshift 'flyswatter' on the manipulator arm in the hope of activating the dead Leasat 3 satellite. Jeffrey Hoffman, shown here, is one of the two astronauts who makes an unscheduled spacewalk to attach the swatter. However, this time Yankee ingenuity doesn't pay off. Leasat stays dead.

▶ (opposite left) **Fixing Leasat**
The shuttle returns to Leasat on mission 51-I to attempt repairs. Here William Fisher fixes a protective cover over the kick-motor nozzle before he begins the delicate work of installing a new electronics box.

▶ (opposite right) **Job done**
Seen perched on the cherry picker is James van Hoften, who has just manually relaunched Leasat, justifying his nickname of 'the Ox'. The on-board astronauts take this picture through one of the overhead windows of the aft crew station at the rear of *Discovery*'s flight deck.

but its rocket motor refused to ignite. And the 6-meter (20-foot) long, 7-tonne satellite was left stranded 35,400 km (22,000 miles) below its working orbit. Hughes engineers reckoned the problem might have occurred because an arming pin on the side of the satellite had not been triggered.

NASA decided to try to flip the switch with *Discovery*'s manipulator arm. But first something had to be attached to it to snag the pin. This resulted in the famous 'flyswatter' device, which was made from plastic notebook covers. David Griggs and Alan Hoffman then went on EVA to attach the flyswatter, and Rhea Seddon with the manipulator arm tried swatting to catch the pin. But it was all to no avail. It was a potentially dangerous situation too because, had Leasat's motor accidentally been ignited, *Discovery* could have been severely damaged. The operation was soon called off. The unhappy mission came to a close with a bang, when one of *Discovery*'s tires blew when landing at Kennedy.

The Ox at work

With the successful recovery of Palapa and Westar very much in mind, NASA decided to mount a rescue mission for Leasat 3. This mission (51-I), again with *Discovery*, blasted off on 27 August. The mission specialists involved this time were spacewalkers William Fisher and James van Hoften, nicknamed 'the Ox' because of his exceptional strength. Operating the manipulator arm was John (Mike) Lounge.

After the successful deployment of three satellites, *Discovery* on 1 September maneuvered to within 11 meters (35 feet) of the dead Leasat. Then van Hoften hitched a ride on the manipulator arm to the slowly spinning satellite. He attached a grappling bar to it and gave it a shove to stop its rotation. Then the arm carried him and Leasat into the payload bay.

Fisher (a medical doctor as well as an engineer) then took over to perform the necessary surgery on the satellite and install a new

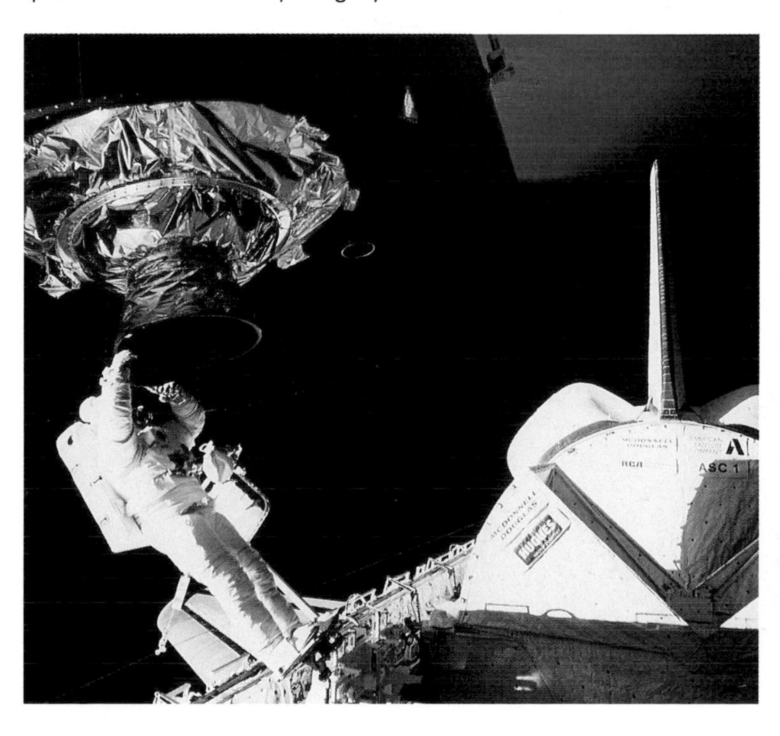

electronics module designed to by-pass the faulty timing mechanism that had made Leasat fail the first time. They completed most of the work that day. The next day Fisher finished off and it was left to van Hoften to relaunch Leasat. He first set it slowly spinning and then shoved it away. 'There that bad boy goes,' he said.

EASE ACCESS

The success of the Leasat repair – it was soon responding to ground control again – prompted NASA spokesman Jesse Moore to say: 'This repair demonstrated the value of sending people into space, and I hope that reminds everybody of the shuttle's capabilities.'

These capabilities were once more displayed on new orbiter *Atlantis*'s second msision into the heavens (61-B), beginning 27 November 1985. During this mission, NASA (as always!) dubbed it 'the best flight ever'. The main reason was the work done during two spectacular EVAs by Sherwood Spring and Jerry Ross. Working in the payload bay, they built up trussed structures similar to those that will in a few years time be used to build the foundations of space station *Freedom* (see page 200).

Two different methods were used to build the structures. In the EASE (experimental assembly of structures in extravehicular activity) method, the two astronauts carried out the assembly of short beams and struts from fixed work stations. In the ACCESS (assembly concept for construction of erectable space structures) method, the astronauts moved about, riding the manipulator arm, and carried out the assembly using a few large beams with interconnecting struts.

Saving Skylab

Spacewalking construction-engineer astronauts will in the future be needed not only in space-station construction. They will also need to be on call at all times when the station becomes operational. This much has been learned already from the experience gained with past space stations, such as Skylab and Salyut (see page 188). The early Salyut missions included no plans for EVA, unlike Skylab, where the long-stay astronauts were scheduled to make regular sorties to the outside to retrieve film canisters, for example.

In the event Skylab astronauts were forced to begin spacewalking much earlier – in fact at the very beginning of the mission. During the launch of the unmanned station on 14 May 1973, part of its micrometeoroid/heat shield was ripped away. So was one of the solar panels; and another panel became jammed. Immediately on reaching orbit, Skylab began to overheat as the Sun penetrated the damaged area. And it was desperately short of electrical power because of the disaster with the solar panels.

Initially it seemed that the whole $2.5 billion project would have to be abandoned. However, NASA engineers reckoned that, with a bit of luck, the space station might, just might, be saved by on-the-spot repairs by a spacewalking crew. Nothing like this had ever been done before, but it was worth a try.

So on 25 May, 10 days later than originally planned, Charles Conrad, Joseph Kerwin and Paul Weitz sped into orbit and rendezvoused with the ailing space station. They took with them a variety of tools, including wire cutters, together with a sunshade, or parasol, to rig over the damaged area. As they drew close to Skylab, they confirmed that the situation was much as expected. Reported Conrad confidently to Mission Control: 'I think we can take care of it.' But he was a little premature. Their attempts, by means of a lean-out EVA from their Apollo spacecraft, ended amid audible obscenities in failure.

They then decided to attack the problem from the inside. They docked with Skylab with difficulty, and then found the temperature inside to be almost unbearable – over 50°C. The most urgent task was obviously to erect the parasol. This they did, working through the airlock. Within hours, the temperature began to fall. After a few days, they were able to take up residence. On the seventh day of the mission Conrad and Kerwin went on a potentially hazardous EVA to free the jammed solar panel. Using a pair of cutters on a long pole, they again achieved success. The panel deployed and soon began feeding much needed extra power into Skylab's electrical system.

These vital spacewalks not only enabled the first crew to complete its planned 28-day mission; they also made Skylab habitable for the other two missions of 59 and then 84 days.

Salyut EVAs

The 84-day space endurance record set by the last Skylab crew, who returned in February 1974, stood for four years until it was beaten by two cosmonauts in Salyut 6. They were Yuri Romanenko and Georgi Grechko. And it was these cosmonauts who on 19 December 1977 carried out the first Soviet EVA for nine years. The EVA was necessary to inspect one of the two docking ports of the Salyut for damage. By all accounts it almost ended in disaster when the untethered Romanenko, who should have stayed in the airlock, floated out. He was caught by Grechko just before he drifted out of reach.

Since then EVAs have been performed at regular intervals, some for emergency practice, others for retrieving test samples on the outside of the craft. On occasions the long-stay cosmonauts have been tempted to dally, for good reason. In July 1978, for example, Vladimir Kovalyonok, while spacewalking with Alexander Ivanchenkov, refused to go back in when ordered by ground control, saying: 'We would just like to take our time because it is the first time in 45 days that we have been out of doors for a walk.'

Cosmonauts began making longer and longer missions in Salyut 6 and then in its successor, Salyut 7, which went into orbit in 1982. It was still in orbit, though mothballed, in 1989, the last cosmonauts having visited it in October 1985. During its three years of occupancy, it suffered many systems failures, and EVAs were often required to fix them. In particular, several extra sets of solar panels were fitted to boost the electricity supply. Spacewalks from *Mir,* Salyut's successor, have included one by French cosmonaut Jean-Loup Chretien in December 1988. He became the first non-Russian, non-American to enter raw space.

◀ Svetlana at work
On 25 July 1984 cosmonaut Svetlana Savitskaya became the first woman to go spacewalking, from the space station Salyut 7. She was, again a record for a woman, making her second appearance in space. She had made her space debut nearly two years before, on 19 August 1982. During her spacewalk, which lasted for 3½ hours, she and fellow-walker Vladimir Dzhanibekov practiced using an electron-beam welding and cutting tool.

Chapter 6

WALKING ON THE MOON

◀ **Apollo 11: Lift-off**
The five engines of the mighty Saturn V Moon rocket erupt into flame and start to lift the 3000-tonne monster off the launch pad. The date is 16 July 1969.

▶ **Apollo 17: EVA**
The desolate and hauntingly beautiful landscape near the landing site in Taurus-Littrow valley, where man explored the Moon for the last time this century. Inset is a possible scene on the Moon next century.

Just right of center on the Moon's surface, as we view it from the Earth, is a dark circular patch we call the Sea of Tranquillity. It was near the western edge of this sea that a human being from planet Earth first stepped down on to another world. The date was 20 July 1969; the human being, American civilian astronaut Neil Armstrong; his spaceship Apollo 11.

Between that date and 14 December 1972, the date of the last Apollo mission, a total of 12 astronauts left their footprints in the lunar soil in six spectacular voyages of discovery. These trailblazing lunarnauts roamed and rode across the great lunar lava plains and rugged highlands for a total of 80 hours. They carried out experiments, set up scientific stations, collected 385 kg (850 pounds) of Moon rocks and soil, acquired 20,000 reels of taped data and took more than 30,000 photographs.

To the lunar scientist the samples and data were treasures from the heavens beyond compare. But to the layman it was the stunning photographs of alien landscapes that fired the imagination, as they do still. They show the Moon to be a place of ethereal beauty, of 'magnificent desolation'. They also reveal the beauty of our own planet, an oasis of life, color and warmth in the stark black desert that is space.

▲ **Agena target**
Docking after rendezvous is another vital technique. Here Gemini 12 astronauts have already docked with their target and attached a tether.

▼ **Gemini 7**
The capability for craft to rendezvous in space is vital to the method chosen to take men to the Moon. Spacecraft rendezvous in orbit for the first time in December 1965. Gemini 6 astronauts take this picture of their sister craft, keeping station a scant few meters away. The cloud tops lie 260 km (160 miles) below.

THE JOURNEY that led to Tranquillity Base, where Neil Armstrong first set foot on the Moon in 1969, began at a leisurely pace more than a decade earlier. The pace quickened into a sprint after President Kennedy's famous speech to Congress in May 1961 urging the American people to aim for the Moon with the utmost urgency: to land a man on the surface and return him safely to Earth before the decade was out (see page 44).

NASA took up the President's challenge and drew up plans for a lunar landing within the time limit imposed. These plans involved a three-stage strategy. The first requirement was actually to launch Americans into orbit, which was the objective of Project Mercury. The second was to broaden their experience of the space environment and practice the skills that would be required to achieve a lunar landing, which was the objective of the Gemini program. All these objectives satisfied, the actual Moon landings could then go ahead under the Apollo program. For each stage new technologies would have to be developed and ingenious and often gargantuan hardware built.

The method NASA selected to effect a lunar landing was called lunar orbit rendezvous. It demanded a three-module spacecraft carrying three astronauts – a command module (CM) to house the crew; a service module (SM) to house equipment and a rocket motor; and a lunar module (LM), in which two of the astronauts would carry out the actual Moon landing. For most of the journey to and from the Moon, the command and service modules would be joined together as the CSM, or mother ship. The LM would be docked with the CSM for the outward journey, separating in lunar orbit prior to the lunar landing. The crew would return to Earth in the CSM and land in the CM.

Paving the way

First the one-man Mercury flights and then the two-man Gemini flights were successfully concluded. The astronauts mastered the art of rendezvousing and docking in space, so essential to the Moon-landing concept. They also showed that spacewalking

▼ **Lunar landscape**
High-resolution pictures like this, taken by Lunar Orbiter probes orbiting the Moon, help NASA decide on suitable sites for the Apollo landings. Later, close-up observations by human pathfinders will clinch their decisions.

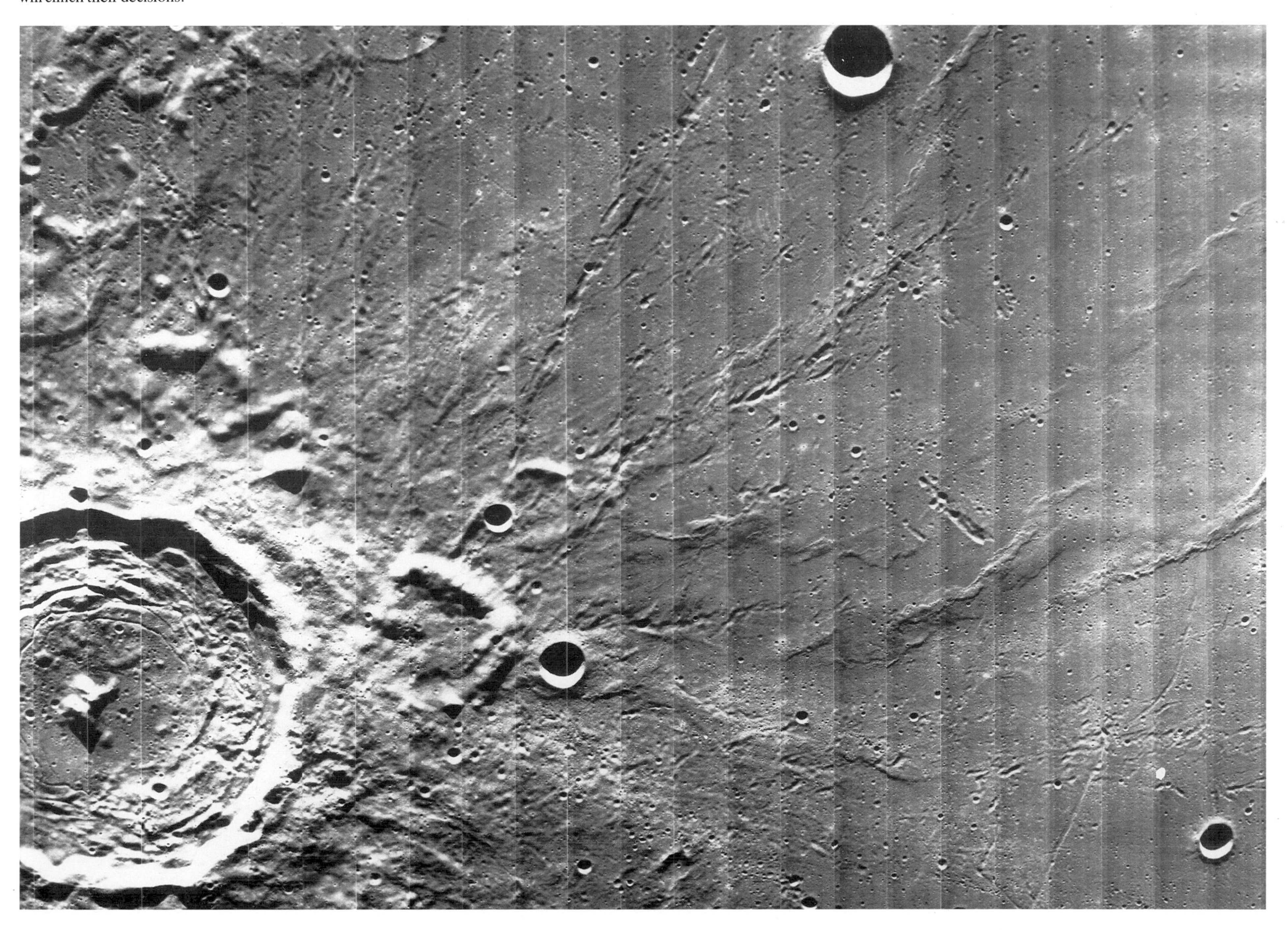

◀ Saturn V on the pad
The Moon rocket and its target. The 36-storey high Saturn V sits on the pad. Its successful maiden flight on 9 November 1967 augurs well for the future. Packing more than 3.5 million kg (7 million pounds) of thrust, it is the mightiest rocket there has ever been.

▶ (far) **Apollo 12: LM descent**
Two of the three astronauts in the Apollo spacecraft fly the lunar module down to the Moon's surface. This classic Apollo photograph shows the lunar module just after separation from the CSM during the second Moon-landing mission in November 1969. What stark beauty the Moon has.

▶ LM touchdown
As the lunar module descends from orbit, it speeds up under the attraction of lunar gravity. To slow them down for a soft landing, the astronauts fire the descent engine as a brake. They land gently on the shock-absorbing legs.

presented no fundamental problems. The Gemini program had achieved its objectives – and more – by the time it reached its conclusion in November 1966. And by then other elements in the run-up to the Apollo missions were beginning to knit together. Dummy Apollo spacecraft were being successfully flown. The real spacecraft were on the production line. Unmanned probes – Ranger, Lunar Orbiter and Surveyor – were being dispatched on pathfinding missions to the Moon. Their task was to take detailed photographs of the surface from which the Apollo landing sites could be selected.

The Surveyor series of soft-landers not only took close-up pictures of the surface, they also analyzed the soil and tested its structure. The landings themselves proved that the Moon's surface was firm enough to take spacecraft. Previously there had been fears that the whole of the surface might be covered in thick layers of soft dust, into which landing spacecraft would sink. Happily, the Surveyor landings proved otherwise.

The first manned Apollo mission, Apollo 1, was scheduled for February 1967. But it was not to be. A month before they were due to take off, the prime crew of Virgil Grissom, Edward White and Roger Chaffee perished in a flash fire while they were in a practice countdown in their command module on the launch pad. They became the first martyrs in the American space program. More than a year went by before the first manned Apollo mission, Apollo 7, with a redesigned command module, blasted off into space, in

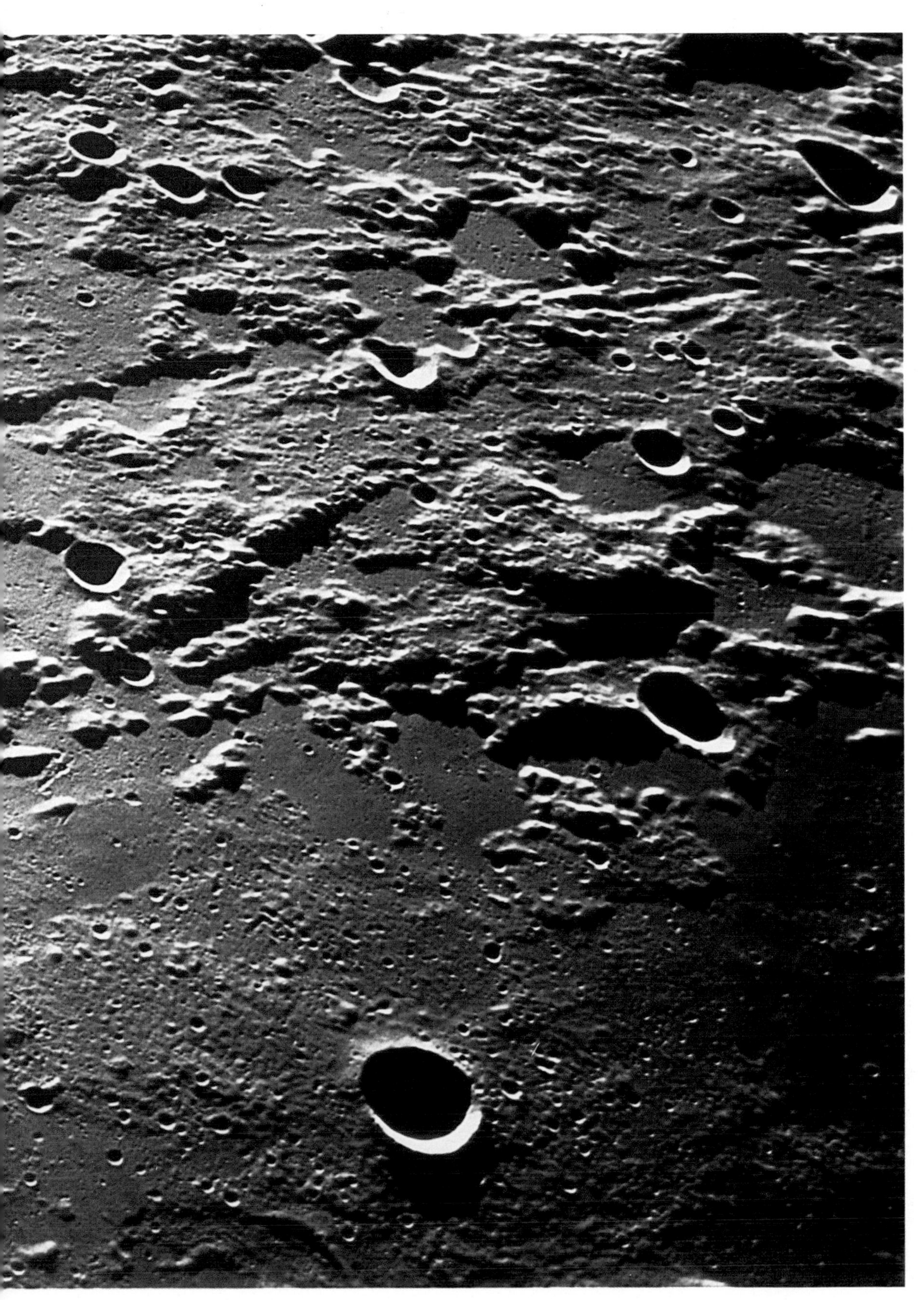

October 1968. Three more flights took place to pave the way for the greatest adventure of all time – the first Moon landing. In December 1968 Apollo 8 set out to circumnavigate the Moon, 'man-rating' the Saturn V and refining translunar navigational parameters.

In March 1969 Apollo 9 test flew the entire Apollo spacecraft – CSM and LM – for the first time. Two months later Apollo 10 flew to the Moon to perform a full dress rehearsal of the lunar landing, without actually landing. The two astronauts in the LM swooped within 15 km (9 miles) of the lunar surface. How great must have been the temptation to go the whole hog and land. But they didn't. They stuck to their role as pathfinders for the momentous event that was to follow.

'The *Eagle* has wings'

That event, the historic mission of Apollo 11, began on 16 July 1969 as Florida was gripped in high summer. Lift-off took place at 9.32 am Eastern Daylight Time (EDT). A million people, crowded into the Kennedy Space Center and every vantage point around Titusville, Cape Canaveral and Cocoa Beach willed the 36-storey high Saturn V rocket into the heavens. Kennedy Launch Control wished 'Good Luck and Godspeed' to the crew of Neil Armstrong, Edward Aldrin and Michael Collins.

Within a quarter of an hour, Apollo 11 was in orbit 190 km (120 miles) high, with the third-stage rocket still attached. Shortly after midday this rocket burst into life once more and boosted Apollo 11 to a speed of 38,700 km/h (24,200 mph). This kicked the craft into a trajectory that would take it to the Moon four days hence.

The astronauts now had to reconfigure the spacecraft for the 385,000-km (240,000-mile) flight. They maneuvered the CSM (callsign *Columbia*), away from the third stage, turned round and edged back to dock with the lunar module *Eagle* inside. That accomplished, they fired the CSM's thrusters and drew the whole spacecraft away from the third stage.

By 11 pm EDT on 18 July, Earth's gravity had reduced Apollo 11's speed to a mere 3300 km/h (2060 mph). But then, just 62,800 km (39,000 miles) from its target, it began to accelerate once more as it came under the gravitational influence of the Moon. A little before 1.30 pm on 19 July the crew fired *Columbia*'s engine as a brake so that it could be captured by lunar gravity and enter orbit. Next morning, 20 July, Aldrin and Armstrong crawled into *Eagle* and powered it up. At 1.46 pm, while they were still behind the Moon, they separated from *Columbia*, leaving Collins on his own. When, a few minutes later, they came into radio contact again with Mission Control at Houston, Armstrong reported: 'The *Eagle* has wings.'

◀ **Apollo 12: Lunar vista**
While his two moonwalking colleagues are busy exploring the lunar surface on foot, the 'forgotten' CSM pilot circles in orbit above. He has little time to be bored, however, having a heavy workload, which includes photographing future landing sites. Richard Gordon snaps this magnificent lunar vista near the crater Lalande, almost in the center of the lunar nearside.

▶ **Apollo 11: Landing site**
The lunar module *Eagle* on the dusty plain of the Sea of Tranquillity. Edwin Aldrin is shown working at the scientific equipment bay in the descent stage. Later this stage will act as a launch pad for the upper one, which will carry the astronauts back into lunar orbit.

▲ **Apollo 11: Plaque**
Attached to one of *Eagle*'s landing legs, beneath the ladder, is a plaque to commemorate its historic landing. Neil Armstrong reads out the inscription. 'Here men from the planet Earth first set foot on the Moon. July 1969 AD. We came in peace for all mankind.' The plaque bears the signatures of Neil A. Armstrong, Michael Collins, and Edwin E. Aldrin Jr, astronauts, and Richard Nixon, President of the United States of America.

Down to Tranquillity

Two hours later they fired *Eagle*'s descent engine and began to drop from orbit. Again, they were out of radio contact behind the Moon. Not until *Columbia*, which had its own radio links with *Eagle*, emerged over the lunar horizon, did an anxious Mission Control know what was happening at this critical time. The news was reassuring. Said *Columbia*'s lone occupant Collins: 'Everything's going just swimmingly. Beautiful!'

Just after 4 pm Armstrong throttled up the descent engine in preparation for *Eagle*'s final approach. The automatic landing system, however, was taking *Eagle* towards a crater the size of a football field, strewn with large rocks. Armstrong took over manual control. His heart-beat leapt from the normal 77 to 156. Soon they had descended low enough for the exhaust from the descent engine to kick up clouds of dust.

As Armstrong searched for a safe landing site, *Eagle*'s fuel began to run low. Mission Control updated him on the flying time he had left, while Aldrin fed him altitude readings. This was the tense dialog that ensued:

Mission Control: '60 seconds.'

Aldrin: 'Lights on. Down 2½. Forward. Forward. Good. 40 feet. Down 2½. Picking up some dust. 2½ down. Faint shadow. 4 forward. 4 forward. Drifting to the right a little.'

Mission Control: '30 seconds.'

Aldrin: 'Forward. Drifting right. Contact light. Okay, engine stop . . . .'

Mission Control: 'We copy you down *Eagle*.'

Armstrong: 'Houston, Tranquillity Base here. The *Eagle* has landed.'

Mission Control: 'Roger, Tranquillity, we copy you on the ground. You've got a bunch of guys about to turn blue. We're breathing again. Thanks a lot!'

The dream lives on

It had been touch and go, but they had made it. *Eagle* touched down on the Moon at 4.18 pm. After resting, the crew prepared for the moment that they and the whole world had been waiting for, when a human being would set foot on another world.

That moment arrived at 10.56 pm EDT on 20 July 1969, when Armstrong planted his left foot in the lunar dust. 'That's one small step for a man,' he said, 'one giant leap for mankind.'

President Kennedy's deadline set eight years previously had been met, with just five months of the decade remaining. Tragically, he did not live to see the day when humankind took its first step into the universe. He was assassinated at Dallas, Texas, on 22 November 1963. But his dream lived on. Armstrong, Aldrin

◀ **Apollo 11: Aldrin**
Perhaps the most famous of all space pictures, Edwin 'Buzz' Aldrin posing for Neil Armstrong. Photographer, lunar module and landing site are reflected in Aldrin's vizor.

▲ **Apollo 11: CSM pilot**
The third member of the crew, Michael Collins, pilot of CSM *Columbia*, who remains in lunar orbit.

and Collins and Apollo 11 fulfilled that dream.

Flying the flag

That Apollo 11 was a flag-waving exercise there was no doubt. And, after setting up an experiment to monitor the solar wind (see page 176), Armstrong and Aldrin broke out the Stars and Stripes on a telescopic flag pole. Its top edge was braced by a spring wire so that it could 'fly' on a world with no wind. Apollo 11 also carried two other American flags, together with the flags of the United Nations and 136 other countries.

Shortly after erecting the flag, history's first two lunarnauts were informed by Mission Control that the US President, Richard Nixon, was on the phone from the White House. With hundreds of millions of the Earth's peoples eavesdropping, the President eulogized over their feat. 'Neil and Buzz,' he said. 'This certainly has to be the most historic telephone call ever made.... For every American this has to be the proudest day of our lives.... Because of what you have done, the heavens have become part of man's world. As you talk to us from the Sea of Tranquillity, it inspires us to redouble our efforts to bring peace and tranquillity to Earth.... For one priceless moment, in the whole history of man, all the people on this Earth are truly one.'

Afterwards, the two lunarnauts set up two more experiments, a seismic detector to detect 'moonquakes' (see page 176) and a laser reflector to measure accurately the Earth-Moon distance. Then it was time to leave what Aldrin called in wonder when he first stepped down on to the Moon, this 'magnificent desolation'. At 1.11 am on 21 July 1969 the first human beings to leave their footprints on the Moon were back in their spaceship *Eagle*, which had brought them safely down and which was soon to whisk them back to lunar orbit to join *Columbia* and its lonely occupant. Observed Mission Control later that morning before the crew were reunited: 'Not since Adam has any human known such solitude as Mike Collins is experiencing during this 47 minutes of each lunar revolution when he's behind the Moon.'

▲ **Apollo 11: After EVA**
After their triumphant first walk on the Moon, Aldrin (above left) and Armstrong (above) relax in the lunar module. They have with them 22 kg (48 pounds) of Moon rocks and soil.

◀ **Apollo 11: Moonwalkers**
Armstrong and Aldrin pictured together by an automatic camera as they unfurl the American flag.

For posterity

The moonwalkers lifted off the Moon at 1.54 pm on 21 July. As well as the discarded hardware associated with their exploration, they left behind on the surface symbolic mementoes for posterity.

They left a gold olive branch, the traditional symbol for peace. They left the mission emblem of the ill-fated Apollo 1 crew, who perished before they could reach for the stars. They left a memorial for the Soviet cosmonauts who had lost their lives in space accidents. And they left a silicon disc about the size of a 50 cent piece on which were etched, by electronic microminiaturization techniques, messages from world leaders and the Pope. They also included the evocative words of President John F. Kennedy which launched the Apollo program: 'I believe that this Nation should commit itself to achieving the goal, before this decade is out, of landing a man on the Moon and returning him safely to Earth.'

Exploring the lunar lowlands

Apollo 11 set down on the Sea of Tranquillity, a vast featureless plain covered in dust, strewn with rocks and pockmarked with craters large and small. Tranquillity is typical of the lunar seas. The term 'seas' was adopted by ancient astronomers centuries ago. In the dark-colored flat regions and contrasting light-colored rugged regions on the Moon, they saw resemblances to the flat ocean areas and rugged land masses of the Earth. And so they named the dark areas 'maria' (singular 'mare'), or seas, and the light areas 'terrae', or continents. Astronomers still use the Latin names for the seas: the Sea of Tranquillity is Mare Tranquillitatis.

The second lunar landing mission, Apollo 12 (November 1969), which incidentally was struck by lightning during lift-off and suffered a temporary power blackout as a result, also aimed for a lunar sea. The target was a site on the vast Ocean of Storms (Oceanus Procellarum), largest of the Moon's seas. The Apollo 12 lunar module *Intrepid* made a pinpoint landing within 185 meters (600 feet) of the Surveyor 3 lunar probe, which had set down two-and-a-half years previously to take photographs and to examine the nature of the lunar soil.

The first attempt at a third lunar landing (April 1970) by Apollo 13 went drastically wrong two days after lift-off. The CSM was put out of action by an explosion, which almost cost the crew their lives (see page 69). The second attempt, by Apollo 14 (January 1971), was successful and the lunar module *Antares* made lunar landfall near Fra Mauro, a crater in a rolling hilly region at the eastern edge of the Ocean of Storms. The landing site was only about 180 km (110 miles) away from that of Apollo 12. But the landscape at Fra Mauro was quite different from the bland flat mare site of Apollo 12. It was covered with rocky debris probably thrown out when giant meteorites impacted the Moon and created the Sea of Showers (Mare Imbrium) and the prominent craters Copernicus and Eratosthenes.

Into the highlands

The last three missions explored infinitely more dramatic landscapes much farther north and south of the first ones, which were situated close to the lunar equator. Apollo 15 set down in the foothills of the lunar Apennines, a mountain range with peaks soaring to the height of Everest. The landing site was located at the south-eastern edge of the Sea of Showers in a bay with the

◀ **Apollo 16: Full Moon**
We never see this full Moon from the Earth. Most of the left hemisphere shown here is the usually hidden farside. The lunar sea at the center of this image is the Sea of Fertility, which merges at the right with the Sea of Tranquillity, which in turn gives way to the Sea of Serenity and, on the horizon, the Sea of Showers. The circular sea below center is the Sea of Crises.

▼ **Apollo landing sites**
The location of the sites of the six Apollo landings, marked on a photograph of the full Moon taken from Earth. Apollo 11 landed on the Sea of Tranquillity; Apollo 12 on the vast Ocean of Storms; Apollo 14 not far away at Fra Mauro. Apollo 15 and 17 set down in highland areas on opposite sides of the Sea of Serenity; while Apollo 16 landed in highlands much farther south.

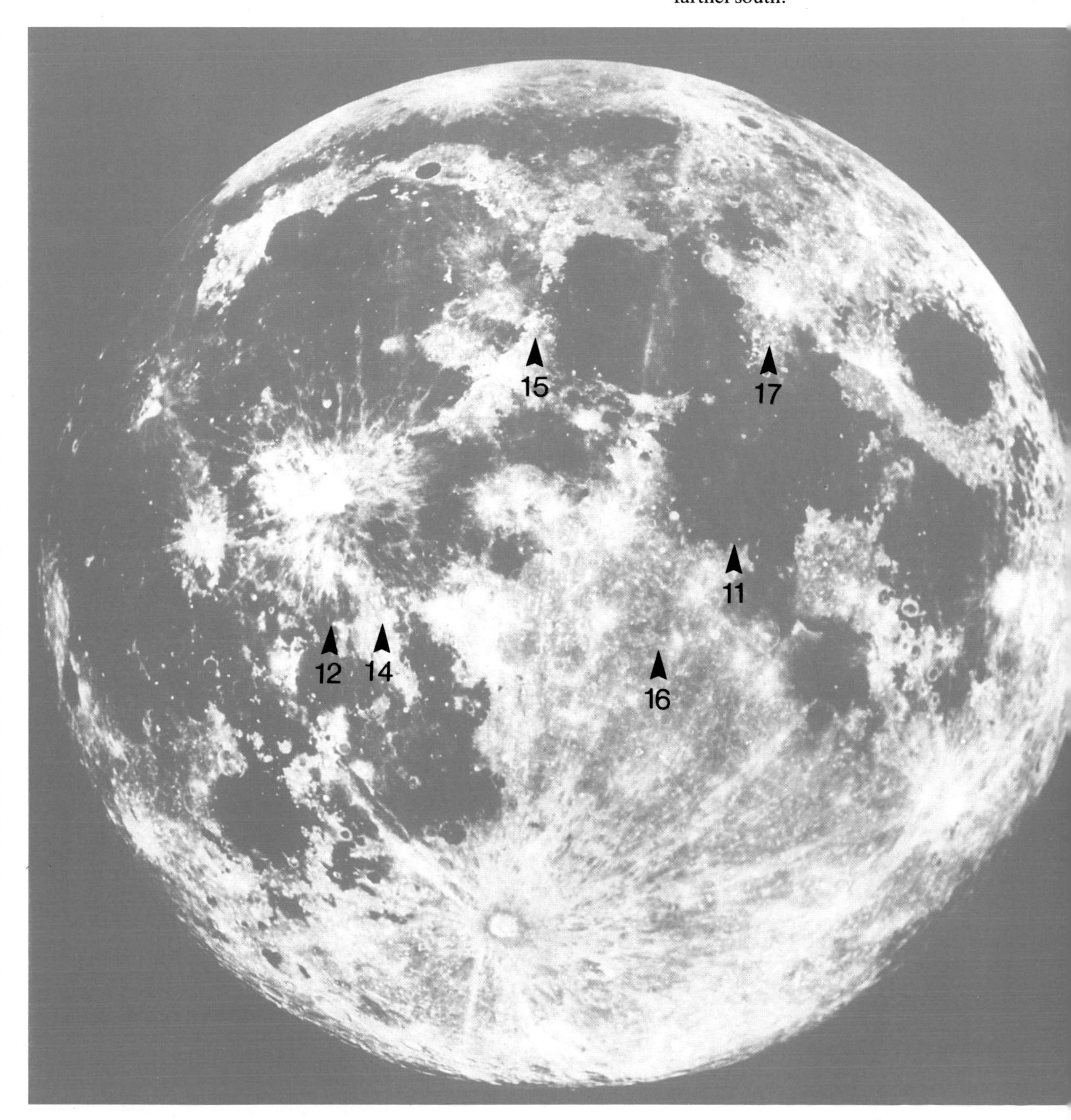

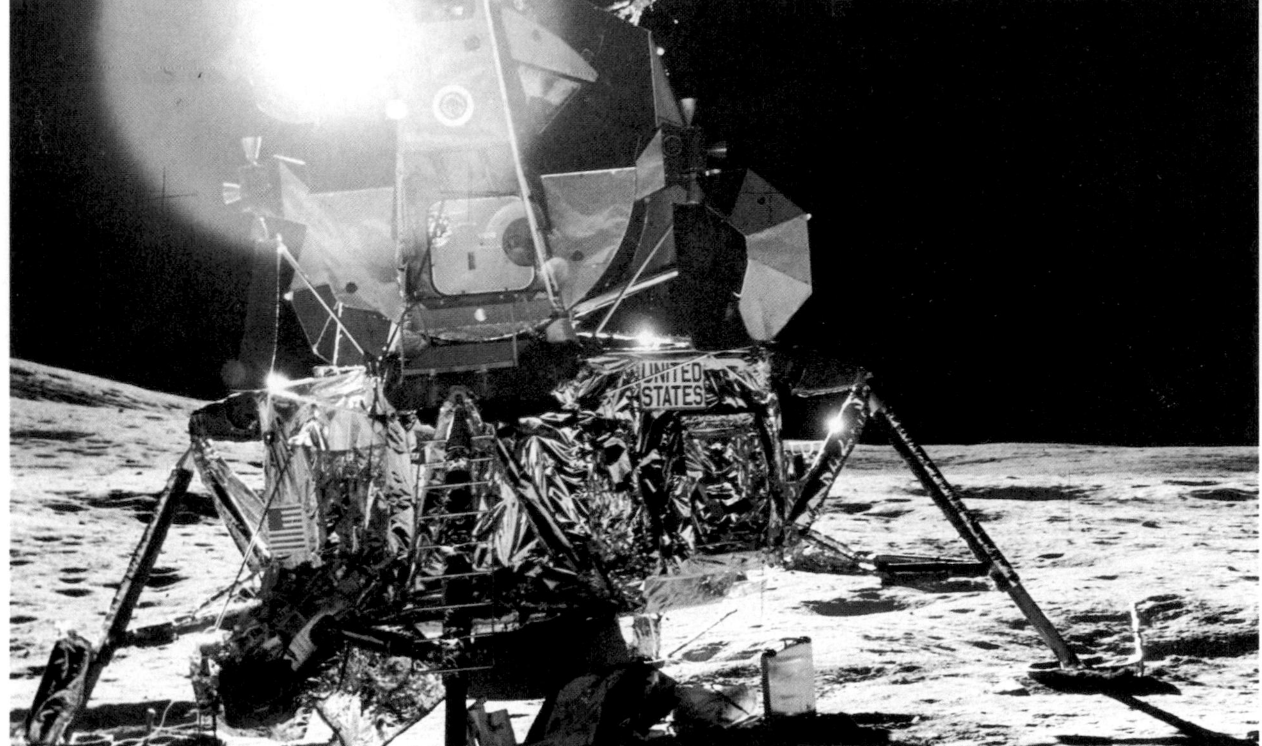

▲ **Apollo 14: LM**
The lunar module *Antares* sits on the surface at Fra Mauro. The strong sunlight of the lunar evening turns the foil-covered descent module pure gold. This view shows the nine-rung descent ladder, and above it the entry hatch into the ascent module.

▶ **Apollo 15: Landing site**
When Apollo 15 visits the Moon, it touches down in the most dramatic landscape yet. This famous picture shows the lunar module *Falcon* against a backcloth of the lunar Apennines in the highland region that separates the Sea of Serenity from the Sea of Showers. Saluting the Stars and Stripes is James Irwin. On the opposite side of *Falcon* is the lunar roving vehicle, about to be used for the first time.

▲ **Apollo 15: SIM bay**
While James Irwin and fellow moonwalker David Scott have been exploring the surface, Alfred Worden has been orbiting above in the CSM *Endeavour*. This shot of the CSM was taken from *Falcon* as it returned to lunar orbit. Part of the service module is open to space. This part, the scientific instrument module bay, houses instruments and cameras to scan and photograph the lunar surface. On the return journey home Worden will undertake the first deep-space EVA to retrieve film from canisters in the bay.

◀ **Apollo 17: Gardening**
Geologist-astronaut Harrison Schmitt gets down to some serious lunar 'gardening'. He is using a kind of rake to sift the surface soil and collect rock fragments.

▲ **Apollo 11: Soil sampling**
Edwin Aldrin uses a specially designed tool to take a sample of lunar soil. Analysis of the soil and rock he and the other astronauts acquire reveal both similarities and differences between them and the rocks and soil found on Earth.

▲ **Apollo 16: Core sampling**
Near the edge of Plum crater, Charles Duke is boring into the surface to take a core sample. The tube traps a long core, which shows the layers of dust that have accumulated over the years.

forbidding name of the Marsh of Decay (Palus Putredinis). The highest peak visible from the site was the 4600-meter (15,000-feet) high Mount Hadley. The site also boasted a deep snaking gorge known as Hadley rille. All in all it was a fascinating site and sight.

Apollo 16 touched down amidst the rugged peaks of the lunar highlands, in a region called the Cayley Plains, near a large crater called Descartes. The landing site was some 250 km (150 miles) south-west of Tranquillity Base, and some 2400 meters (8000 feet) higher. The final landing mission, Apollo 17, targeted a site in the northern hemisphere known as Taurus-Littrow. It was located at the south-eastern edge of the Sea of Serenity (Mare Serenitatis), just south of the Taurus mountains and near the large Littrow crater. It was in a wide valley hemmed in by steep massifs up to 2100 meters (7000 feet) high. The site was chosen because it appeared to be geologically a particularly interesting area, where there should be examples of old and new crust; and there was photographic evidence of fresh lava flows.

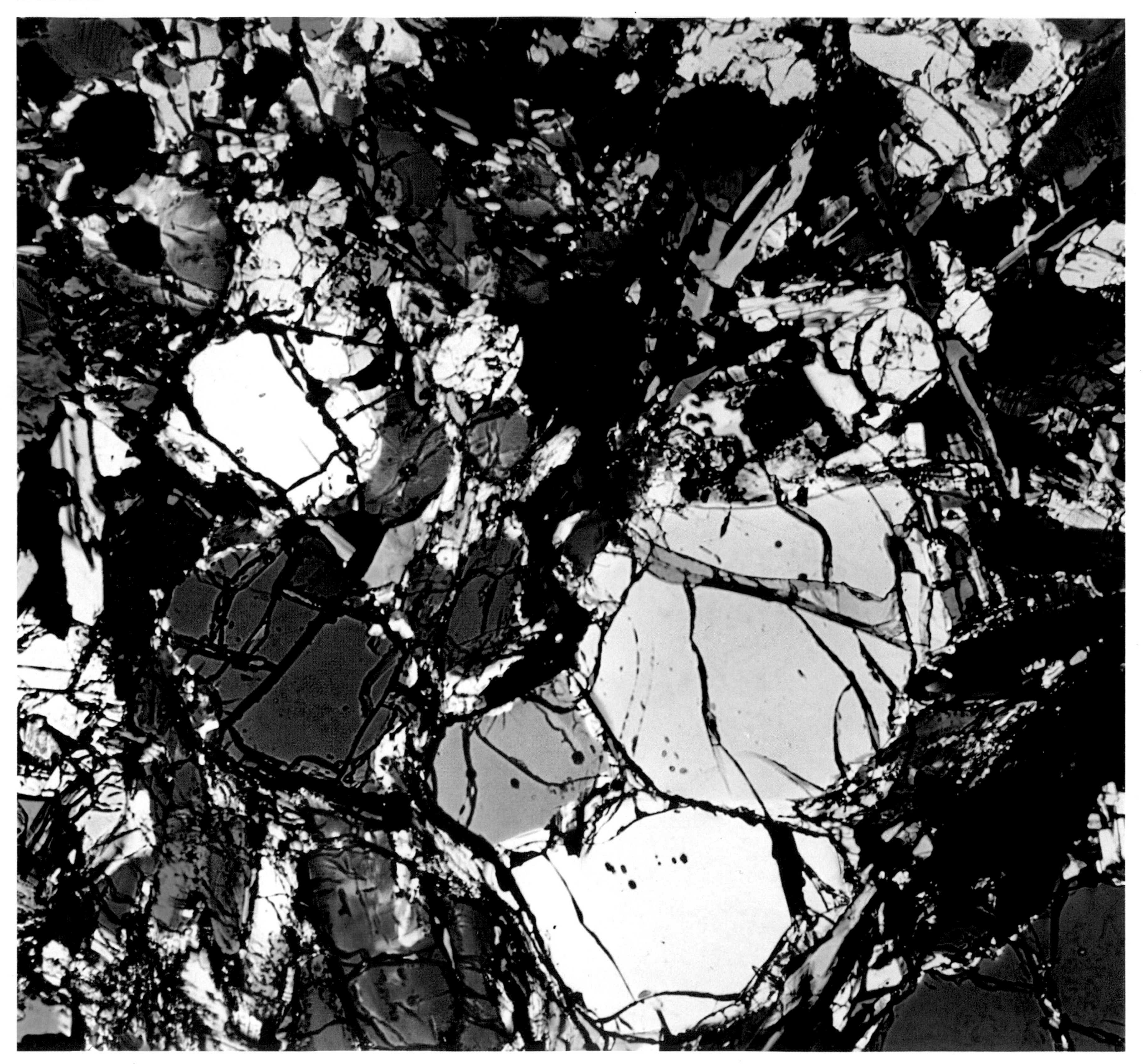

Puffs of Moondust

There was some concern in the early days of the Apollo program that the Moon might be covered with deep drifts of dust that would make landing a spacecraft impossible. These fears proved unfounded, but there is dust everywhere. Immediately after Neil Armstrong took his historic first step, he reported: 'The surface is fine and powdery. I can pick it up loosely with my toe. It does adhere in fine layers like powdered charcoal to the sole and sides of my boots.'

The dust covered everything the astronauts handled on the Moon – and themselves. Some pictures showed them surrounded by a dust haze. When Apollo 17 astronaut Eugene Cernan accidentally knocked off a fender from one of the front wheels of the lunar roving vehicle, this presented a problem because of the dust. Without the fender, the wheels would spray dust over the astronauts and everything else. Fortunately, a make-shift repair cured the problem. Dust kicked up by the descending lunar modules partly obscured the pilots' view prior to landing and forced some of them to land on instruments. Little puffs of dust erupted each time the astronauts took a step. Because of the low lunar gravity, one-sixth of Earth's, the dust settled slowly.

The low gravity also forced the astronauts to modify the way they moved around. But it presented no real problem. As Neil Armstrong first reported: 'There seems to be no difficulty in moving around as we suspected. It's even perhaps easier than the simulations at one-sixth g that we performed in the simulators on the ground.'

Doing the kangaroo hop

Later Edwin Aldrin advised: 'You do have to be rather careful to keep track of where your center of mass is. Sometimes it takes about two or three paces to make sure you've got your feet underneath you. About two to three, maybe four easy paces can bring you to a nearly smooth stop.' They demonstrated the kind of gait scientists had recommended, the 'kangaroo lope', a kind of hopping motion, but found this little more effective than the one-foot-in-front-of-the-other method favored by the Earthbound. The point Aldrin made about the center of mass related mainly to the bulk of their spacesuit and backpack – their portable life-support system. The backpack on Earth weighed 86 kg (190 pounds), a weight comparable with that of the astronauts themselves. On the Moon it weighed only about 14 kg (31 pounds). Although the weight of the backpack was much less, its mass was the same, as was its inertia, and this had to be taken into consideration particularly when changing direction.

◀ **Breccia**
All over the Moon there are chunks of rock like this. It is breccia, made up of cemented rock chips.

◀ **Moon crystals**
In polarized light in a geological microscope, the crystals in a thin slice of Moon rock show up in brilliant colors. This sample of rock came from the Ocean of Storms (Apollo 12).

▼ **Vuggy**
This volcanic rock, riddled with holes like Swiss cheese, came from the vicinity of Hadley rille.

On Armstrong's recommendation, all the astronauts who followed took some time to familiarize themselves with the techniques of moonwalking at the start of their EVAs. Apollo 12 astronauts Alan Bean and Charles Conrad found the novelty of one-sixth g exhilarating and cavorted around uttering whoops of delight. 'Boy, you sure lean forward,' observed Bean. 'Don't think you're going to steam around here quite as fast as you thought you were,' rejoined Conrad.

Occasionally when fatigue set in or when the astronauts became overly absorbed in what they were doing, they took a tumble. During the Apollo 15 mission to Hadley rille, David Scott tripped and fell as he was edging down the slope of the trough-like rille. Had the slope been steeper, he could have skidded down the slope in the slippery dust. And if he had snagged his spacesuit on a jagged rock, he would have perished. Fortunately he didn't, and fellow moonwalker James Irwin was on hand to help him to his feet. Commented an embarrassed Scott as he continued on his way: 'This time I'll look and make sure I don't fall over some silly rock!'

Attack from outer space

Another feature of the lunar soil was that it was slippery, which contributed to the astronauts' tendency to tumble if they were not careful. The slipperiness of lunar soil, called the regolith, is caused by the presence of tiny glass beads. These have been formed as a result of the constant bombardment the lunar surface has received from meteorites. Meteorite impact melts surface material, which then splashes about. This molten rock debris quickly cools into glass fragments, which mixes with the other surface material. The glass beads featured prominently in the samples of rock and soil brought back by the astronauts.

The first samples the astronauts took were from the immediate vicinity of their landing site. These were termed contingency samples, acquired at the beginning of the first EVA so that if they had to quit the Moon in a hurry, they still had something to take back. Later they took samples from farther afield and were more selective in what they chose. In so doing they demonstrated how superior is the astronaut to the space probe, which cannot be programed for such selectivity.

▼ **Apollo 11: Seismometer**
This is one of the three experiment packages the astronauts set up on the dusty surface at Tranquillity Base. The sensor is a seismometer, an instrument designed to detect 'moonquakes'. The equipment includes a radio transmitter to relay readings to Earth.

▶ **Apollo 16: Gnomon**
Up on the lunar highlands Charles Duke is working near the lunar roving vehicle. In the foreground is a gnomon, essentially a stick that casts a shadow. It is used as a reference for Sun angle. It also carries a scale for color comparison for visual and photographic reference.

▶ (far) **Apollo 11: Solar wind**
In the harsh glare of the low Sun, Edwin Aldrin stands next to the solar wind experiment. This consists of a sheet of foil designed to collect particles from the 'wind' streaming from the Sun. Later the astronauts will roll up the sheet and take it back to Earth for analysis.

The astronauts picked up a varied selection of rocks from the scattered landing sites. Some were dark, lava-like and riddled with holes. Some were peppered with crystals. Others were made up of light and dark chips. All were found to be volcanic in origin, which was to be expected. On Earth there are not only volcanic rocks, but also rocks that have formed out of the sediment from ancient seas. On the Moon there have been no seas – at least not watery ones – so there are no sedimentary rocks.

Basalt and chips

The typical rock of the lunar seas is basalt, a dark-colored rock made up of tiny crystals. It is similar to the basalt found on Earth, but has a slightly different mineral composition. Basalt is formed on Earth when molten rock erupts at volcanoes and quickly cools as it hits the air. The lunar seas were created when molten lava from the Moon's interior welled up to the surface and filled great basins that had been gouged out by the impact of giant asteroids, or mini-planets. The great lava lakes that formed solidified into the dark dusty plains we see today. Nearly all of the seas are circular and edged with mountains thrown up when the asteroids hit. By radioactive dating of the rocks the Apollo astronauts brought back, the age of the seas has been found to range from about 3.2 to 3.8 billion years old.

▶ **Apollo 15: ALSEP**
John Young stands on the lunar highlands at Descartes surrounded by instruments of the ALSEP. To his left is a drill used to bore into the surface for core samples. To his right is the drill rack and the bore stems.

Many of the rocks brought back from the highlands were found to be older than this, which was again expected. The highlands are what remains of the Moon's original crust, formed before the asteroids hit. Some of the highland rocks were dated at over 4.2 billion years old. This is close to the age of the Moon itself, 4.6 billion years. The typical rock of the highlands is a light-colored volcanic rock similar to a type known on Earth as gabbro.

There is highland rock in the lowlands as well, where it has been thrown by meteorite impact. Another type of rock is found widely in both highland and lowland areas. It is made up of chips of pre-existing rocks, often cemented together by glass. This type of rock is known as breccia, after a similar rock containing rock chips found on Earth. It is more or less a 'frozen' sample of lunar soil, or regolith, which is a mixture of rock chips, glass beads and dust.

As part of their investigation of the Moon's crust, the astronauts bored into the surface in order to take a core sample. They drove a hollow tube into the soil so that a cylindrical section, or core of soil, was forced into the tube, which was then extracted. Core sampling proved one of the most difficult of the astronauts' tasks. Apollo 15 astronauts David Scott and James Irwin together struggled for nearly half an hour to extract and separate the sections of a 2.4-meter (8-foot) long core. Commented Scott: 'Nothing like a little PT to start out the day!'

When drilling into the ground at Taurus-Littrow to put in a probe for a heat-flow experiment, on the final Apollo mission, Eugene Cernan expended so much effort that his heart-beat soared to 150 beats per minute. And doctors at Mission Control, monitoring his physical state as they did with all the astronauts, ordered him to rest. Coming to his aid, Harrison Schmitt tried to extract the drilling tube and fell over in the process. On several other occasions when the astronauts were having a tough time, audible obscenities made their way back to Earth over the radio!

Creating moonquakes

However, the majority of the experiments the astronauts carried out were less demanding physically. They involved setting up equipment to monitor the lunar environment. On the first landing mission the astronauts deployed a seismometer. The purpose of this device was to detect tremors in the Moon's crust, or 'moonquakes'. By studying the way shock waves travel through a body, scientists can build up a picture of how that body is made up. That is how we know what the Earth's layered interior is like.

Seismometers were also deployed at the other landing sites as part of the package of experiments known as ALSEP (Apollo lunar surface experiments package). To test whether the seismometers worked, a number of the third-stage rockets from the Saturn V launch vehicles and some of the discarded lunar modules were deliberately crashed on the Moon. They jiggled the seismometers in a way no one expected. On Earth, ground tremors last for only a few minutes. But on the Moon, they last for up to two hours. As one seismologist put it, the Moon 'rings like a bell'.

From these artificial moonquakes and also moonquakes that occurred naturally, geologists have built up a picture of the Moon's interior. It is composed of layers of different kinds of rock. The top layer seems to be cracked all over, and it is this that seems to be responsible for the Moon 'ringing'. It is possible that the Moon has a molten or semimolten core.

Among the other instruments in the ALSEP were particle detectors, to detect charged particles coming from the Sun (in the solar wind); magnetometers, to measure the Moon's magnetic field; a gravimeter, to detect changes in the Moon's gravity; and an analyzer, to detect gases in the lunar 'atmosphere'. Surprisingly, traces of gases were detected. Some probably came from the lunar interior, others from the soil, which probably absorbs gases from the solar wind.

▼ Apollo 14: ALSEP
Part of the ALSEP station at Fra Mauro. In the foreground is the RTG, the radioisotope thermoelectric generator, which supplies power to the gold foil-covered transmitter in the background.

The ALSEP collection of instruments, placed at all the landing sites except Tranquillity Base, formed automatic scientific stations. They fed their readings to a central transmitter, which radioed them back automatically to Earth. The ALSEP stations had a design life of just 12 months, but were still sending back information years later. Eventually, on 30 September 1977, the collection of ALSEP data was halted through lack of funds. Signals were still being sent out, however, and were later used by the Jet Propulsion Laboratory at Pasadena, California, to help space-probe navigation.

Unscheduled experiments

The Apollo astronauts had a heavy workload, which they carried out for the majority of the time with unrestrained exuberance. On the final Apollo mission Schmitt and Cernan, having finally mastered the extraction of the core tube mentioned earlier, were moved to burst into an improvised duet: 'While strolling on the Moon one day ... in the merry month of December'. Mission Control was not too impressed, informing them that they were way behind schedule.

But occasionally the effort of it all began to tell. On their second moonwalk Apollo 14 astronauts Alan Shepard and Edgar Mitchell found it a struggle climbing up the slope of Cone crater. Shepard, at 47 the oldest man then to have gone into space, began perhaps to show his age. He puffed with the exertion and after kneeling down to pick up a rock had to be picked up himself. He then reckoned that it was too far to walk to the crater rim. Mitchell disagreed: 'Aw, gee whiz,' he said, 'let's give it a whirl.' Shepard objected, but continued. As they climbed farther, both of their heart rates began to soar and Mission Control ordered them to turn back. Snapped Mitchell, annoyed: 'I think you're finks!'

But their ill-humor did not last long. Just before they had to return to the lunar module, Shepard pulled from his spacesuit two golf balls he had smuggled on to the Moon. Taking a tool from his handcart (which NASA termed the modularized equipment transporter), he swung it like a golf club at the first ball. He kicked up dust but missed the ball. But on his next attempt he hit the ball for 'miles and miles and miles' – in practice probably a couple of hundred meters – in the low lunar gravity.

David Scott performed a more meaningful experiment at the Hadley landing site. He tested Galileo's theory that every body, heavy or light, falls to the ground at the same rate. On the Earth air resistance comes into play to confuse the issue. What better place

▶ **Apollo 17: LRV**
Eugene Cernan can't wait to put the lunar rover through its paces at the start of the first EVA. Notice the way the front and rear wheels turn in opposite directions.

▶ (far) **Apollo 17: LRV**
High massifs provide a backcloth to this evocative picture of the lunar rover, with Eugene Cernan about to climb aboard. It is now fitted with TV camera and antennae and is ready for the off. Cernan and fellow moonwalker Harrison Schmitt use the vehicle to good effect, traveling for 35 km (22 miles) on three EVAs totaling over 22 hours. They collect 115 kg (250 pounds) of rock and soil samples.

▲ **Apollo 17: Full Earth**
Of all the pictures of Earth taken during the Apollo missions this is by far the best. The astronauts took it while they were headed for the Moon. The main landmasses visible are Africa and Saudi Arabia at top and the frozen wastes of Antarctica at bottom.

than the airless Moon to test the theory. So Scott held in one hand a geological hammer and in the other a feather. Then he dropped them from the same height at the same time. They hit the ground simultaneously. 'How about that,' said Scott, 'Mr Galileo was correct.'

Riding the lunar hot-rod

On the last three Apollo misions, the lunarnauts were able to venture several kilometers from the landing site without putting their lives in jeopardy. For they had with them a remarkable car to transport themselves and their equipment. Officially called the lunar roving vehicle (LVR), it was called Rover by the astronauts and popularly nicknamed the Moon buggy. Indeed it did look rather like a stripped-down dune buggy.

When the first rover was driven by the Apollo 15 astronauts in July 197l, it was not the first wheeled vehicle to roll over the lunar landscape. The Soviet's remote-controlled Lunokhod 1 ('Moon-stroller') took that prize, landing on the Sea of Showers in November 1970.

The rover was an ingenious vehicle some 3 meters (10 feet) long. It was a jumble of aluminum struts, tubes and wire, which was stowed folded in the side of the lunar module. Pulling a lanyard caused the rover to spring out and erect itself, rather like a foldaway bed. Each one cost in the region of $13 million.

The rover was powered by ¼-horsepower battery-powered electric motors on each of its four wire-mesh wheels. It had neither a steering wheel nor a brake pedal. It was controlled by a single

▶ **Apollo 11: Earthrise**
The astronauts snap the colorful orb of the Earth as it rises over the lunar horizon just after they enter orbit in July 1969. What a contrast it makes with the dark lunar landscape below. Even at a distance of 385,000 km (240,000 miles), the continents show up plainly in gaps between the clouds.

▼ **Apollo 17: Schmitt**
Another classic Apollo photograph, which captures Harrison Schmitt, the Stars and Stripes, and the planet whence they came.

◀ **Apollo 15: LM lift-off**
The lunar module *Falcon* puts on a fireworks display as it quits the Moon on 2 August 1971 at the end of the fourth successful Moon-landing. The lunar rover's remote-controlled TV camera takes this fascinating picture, recording the kaleidoscopic effect caused by the flames from *Falcon*'s ascent-stage engine.

'joystick', rather like that of an airplane, between the two seats. Pushing the stick forwards caused the rover to accelerate. Pushing it sideways turned both sets of wheels, in opposite directions, so as to execute sharp turns. The astronaut who first gave the rover a test drive on the Moon, Apollo 15's David Scott, said it handled well, but needed seatbelts because it gave a 'rock-and-roll ride'. The rover's top speed was a modest 16 km/h (10 mph).

The rover carried a television camera that was remotely controlled from Earth, and this enabled, for example, the spectacular lift-off of the lunar module to be watched when the astronauts left the surface to return to orbit. A gold umbrella-like device mounted on the vehicle was one of two antennae used for communications between the astronauts and Earth and the CSM orbiting above. Exploration out of sight of base amid the confusing lunar landscape presented no problem, for the rover was equipped with a computerized inertial navigation system. Together the three rovers of Apollo 15 through 17 traveled a distance of 90 km (56 miles).

Going home

When the astronauts finished their moonwalking, their mission was far from over. They had to get their treasure trove of Moon rocks, film and data back to Earth. There were still many things that could go wrong. First they had to blast off the Moon in the upper, ascent stage of the lunar module. Then they rendezvoused and docked with the CSM mother ship, which had been orbiting above them all the while they had been working on the ground. In orbit they transferred themselves and their precious samples to the CSM – with as little dust as possible, which was not easy.. The ascent stage was then jettisoned, often being crashed on to the surface to jiggle the seismometers (see page 176).

Then the astronauts fired the CSM's engine to blast them out of lunar orbit and into a trajectory that would carry them three days hence to the Earth. As they neared the Earth, they were accelerated by gravity to a speed of nearly 40,000 km/h (25,000 mph). It was essential that they hit the atmosphere at exactly the right angle. If they didn't, then they were dead. If they went in too steeply, they would experience such extreme re-entry heating that they would burn up. If they went in at too shallow an angle, they would bounce off the atmosphere and disappear into the depths of space beyond all human assistance.

Fortunately, on all the Apollo missions the angle of re-entry was exactly right. Just before re-entry the service module was jettisoned. As the command module slammed into the air, its heat shield glowed red-hot and began to boil away. But it did its job well, protecting the module and the crew inside. The air rapidly slowed

◀ **Apollo 14: Recovery**
The command module has a rather moth-eaten look about it after splashdown on 9 February 1971. The astronauts in the life-raft wear masks and protection overalls for the last time.

down the module until it was traveling slowly enough for parachutes to open and lower it to a gentle splashdown at sea. This tiny module, just a few meters high and wide, was all that remained of the 36-storey Moon rocket that had blasted off just a few days before.

The last command module to return from the Moon, Apollo 17, splashed down on 19 December 1972. This brought to an end one of the most remarkable episodes in the history of humankind, in which 27 Earthlings circled around the Moon, 12 actually left their footprints in the lunar soil, and three escaped death by a hair's breadth.

A plaque on the descent stage of the Apollo 17 lunar module marks where the first chapter in the history of lunar exploration ended. It reads: 'Here man completed the first exploration of the Moon. December 1972, AD. May the spirit in which we came be reflected in the lives of all mankind.'

▼ **Apollo 15: Splashdown**
The three main parachutes lower the scorched command module into the Pacific Ocean five days after *Falcon*'s spectacular lift-off from the lunar surface.

Chapter 7

SALYUT TO FREEDOM

◀ ***Freedom* in orbit**
By the end of the century the international space station *Freedom* will look something like this. The crew will live and work in the cylindrical modules. The shuttle will pay regular visits.

▶ **Skylab in orbit**
Launched in May 1973, Skylab starts with a near disaster and ends in triumph. Altogether nine astronauts spend a total of 171 days experimenting in space, smashing all previous space-endurance records.

On 21 December 1988 a blackened bell-shaped module dropped from the skies above Kazakhstan in Central Russia, parachutes billowing above it. Two of the three cosmonauts inside had not set foot on Earth for a year. It was the finale of the latest Soviet space-endurance spectacular.

Having lost the race to the Moon in the 1960s, Soviet space supremos concentrated on developing space stations called Salyut, designed for long-term habitation by a succession of crews. Their first attempts in the early 1970s were attended by failure and tragedy. Only in the last few years of that decade did the program really take off. By 1986 the Salyuts had been superseded by *Mir*, and it was in this craft that the record-breaking year-long flight took place.

After winning the Moon race, the US turned its attention to developing the space shuttle. But time, money and hardware were found in 1973 to launch a one-off orbital station, Skylab. It was in Skylab that the effects of prolonged weightlessness first became apparent.

With the shuttle up and running again, priority in the US program has switched to the development of a space station called *Freedom*. It will be an international station, with other nations sharing the costs and the benefits of a permanent foothold in space. It should join *Mir* in orbit in the mid-1990s.

◀ Waiting to launch
The last Saturn V Moon rocket provides the muscle to lift the unmanned Skylab into orbit. The fully equipped space station is built into a rocket that formed the third stage of the Apollo launch vehicle.

AN ORBITING SHISH KEBAB, that is what some Soviet journalists called the first space station. It was launched into orbit on 19 April 1971, 10 years and a week since Yuri Gagarin had pioneered manned space flight. The proper name for the space station was Salyut (meaning Salute). Soviet premier Leonid Brezhnev called it 'an important step in the conquest of space'. A spokesman said that the upcoming Salyut program would feature visits by a succession of crews.

The first crew of three set out only four days later, in the spacecraft that was destined to become the most widely used space ferry – Soyuz. This flight was the 10th by a Soyuz. Although they rendezvoused and docked with Salyut, the crew of the Soyuz 10 did not, for some reason, enter the space station. And after only 5½ hours, they undocked and soon made an emergency return to Earth, landing at night. Whatever went wrong, however, could not have been that serious because in less than two months Soyuz 11 sped into orbit to rendezvous and dock. Aboard were cosmonauts Georgi Dobrovolsky, Vladislav Volkov and Viktor Patseyev. This time the cosmonauts docked successfully and took up residence in Salyut. Enthused Dobrovolsky: 'This place is tremendous. There seems to be no end to it!'

After 13 days in orbit, they passed the US space-endurance record set by the Gemini 7 crew; after 18 days, they surpassed the record set the previous year by their fellow cosmonauts in Soyuz 9. They performed a variety of experiments, made observations of the stars and the Earth, grew crops in their aptly named 'Oasis' greenhouse and hatched frog spawn. They exercised regularly each day with chest expanders and wore elasticated chibis, or 'penguin' suits. These were designed to make them use muscles they could not otherwise exercise in weightlessness. After three weeks the doctors pronounced them in exceptionally good health, if slightly fatigued.

After 24 days, they crawled back into Soyuz for their return to Earth. It should have been in triumph. But when, after an apparently normal landing, the hatch of their descent module was opened, the astronauts appeared to be sleeping. But they were not – they were dead, killed by depressurization.

Apparently, during their explosive separation from the rest of Soyuz during re-entry, a valve had opened and let the air out of the cabin. The crew were not wearing spacesuits and thus stood no chance. The disaster halted the Soviet manned space program for more than two years.

During the intervening period Soyuz was redesigned, and preparations were made to launch a follow-up space station, Salyut 2. This went up on 3 April, but on the 14th it suffered a catastrophic explosion in orbit. Fortunately, no crew had yet flown up to it.

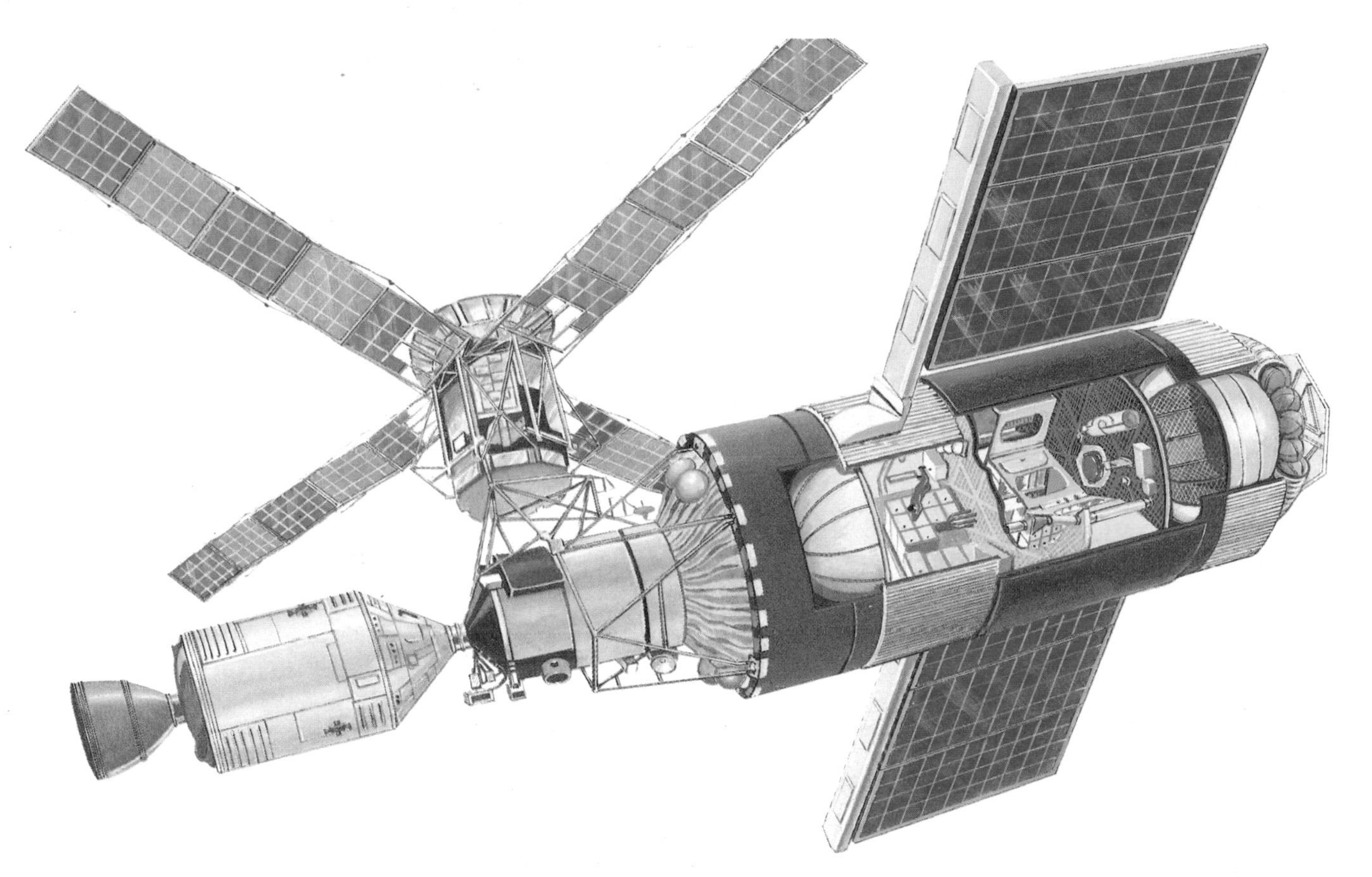

◀ Skylab as planned
This shows how Skylab should have appeared in space. In the event one of the two solar panels on the orbital workshop was torn off at launch. The main accommodation space occupies what was the liquid hydrogen tank of the S-IVB rocket.

▼ Skylab in practice
The space station as seen by the departing final crew of astronauts. A sunshade has been erected over the area damaged when the solar panel was ripped off. Notice there are in fact two sheets of reflective material, put in place by the first and the second visiting crews.

Skylab aloft

The impetus in space-station development now shifted across the Atlantic to the United States, where the countdown was about to begin for the launch of its first space station, Skylab. That launch occurred on schedule on 14 May, with a Saturn V rocket as the launch vehicle. The schedule called for a crew of three in an Apollo spacecraft to be launched next day to Skylab and take up residence there for nearly a month. It would be launched by a Saturn IB. All this hardware was left over from the curtailed Apollo Moon-landing program. The Skylab space station itself was also based on redundant Apollo hardware; it was built around the third rocket stage (S-IVB) of a Saturn V.

The idea of launching a space station predated the Apollo Moon-landing program, for the original Apollo program planned to concentrate first on Earth-orbital activities. Then in 1961 President Kennedy urged the American people to go for the Moon, and the priorities changed. But plans slowly went ahead at the Marshall Space Flight Center at Huntsville and the Johnson Space Center at Houston for an Earth-orbiting station to follow Apollo. The idea was to use as much Apollo hardware as possible to keep down the cost.

The plans centered around using an S-IVB as the base. Huntsville favored a 'wet' approach, launching the S-IVB full of propellants as part of a Saturn IB launch vehicle. When in orbit, the tanks would be drained and then outfitted as a laboratory. Houston favored the 'dry' approach, launching a fully fitted laboratory with a Saturn V. Houston won. And early in 1970 the project, formerly known as the Apollo Applications Program, was renamed Skylab and a 1973 launch was announced.

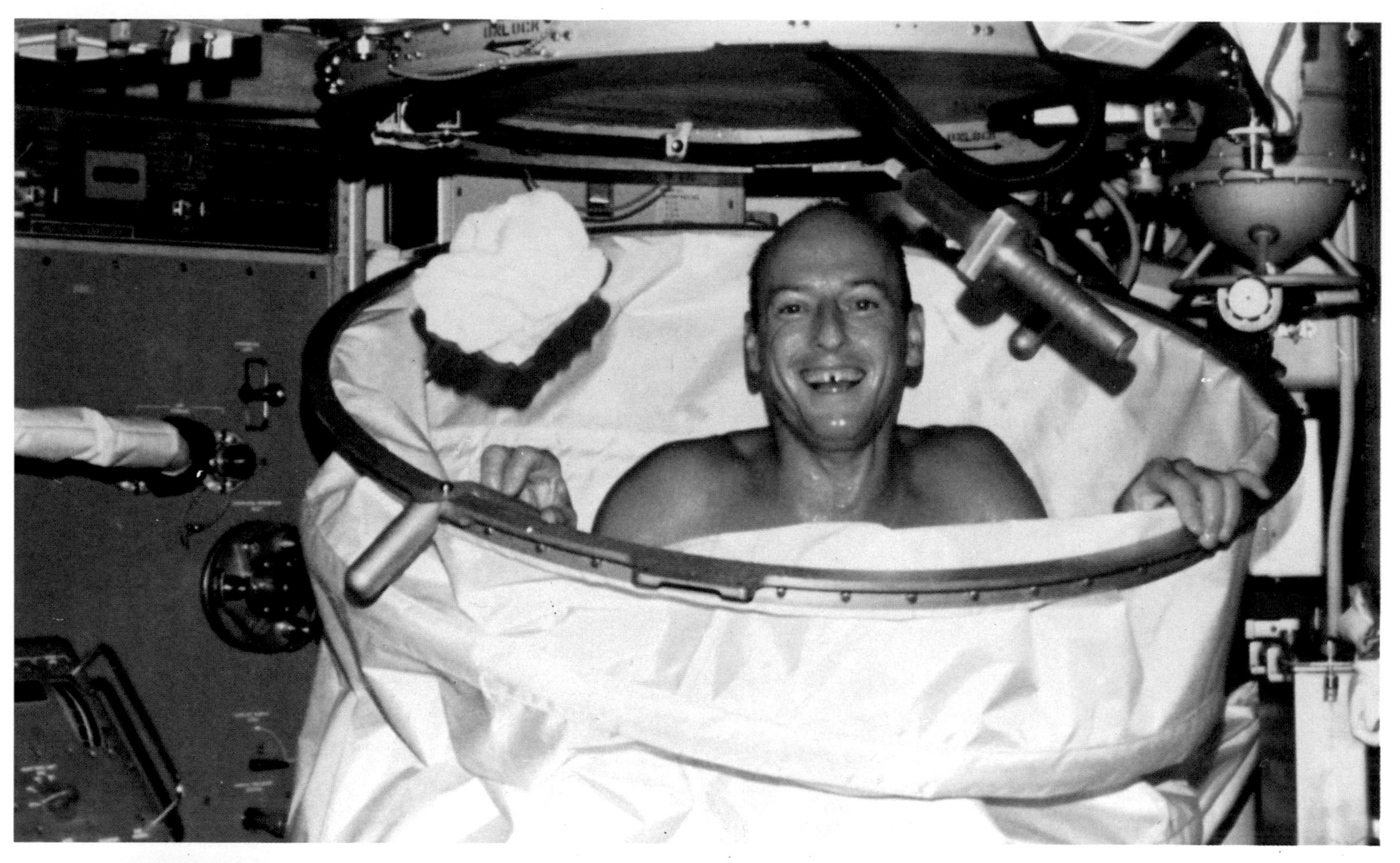

▶ Checking the circulation
Because there is no effective gravity in orbit, the body's circulation, or cardiovascular system, is upset. Here Owen Garriott (Skylab 3) participates in an experiment using a lower-body negative pressure device, which seeks to alleviate the problem.

◀ What a shower!
Skylab 2 visitor Charles Conrad is clearly enjoying the rare luxury of a shower in space. However, water drops escape from the shower cubicle and have to be vacuumed up by his fellow crew members.

◀ Open wide!
Medical checks are carried out regularly on Skylab to monitor the effects of weightlessness on the body. Here Dr Joseph Kerwin gives Conrad's mouth the once-over. This picture confirms what Tsiolkovsky once said about life in space, that anywhere would serve as a chair or bed.

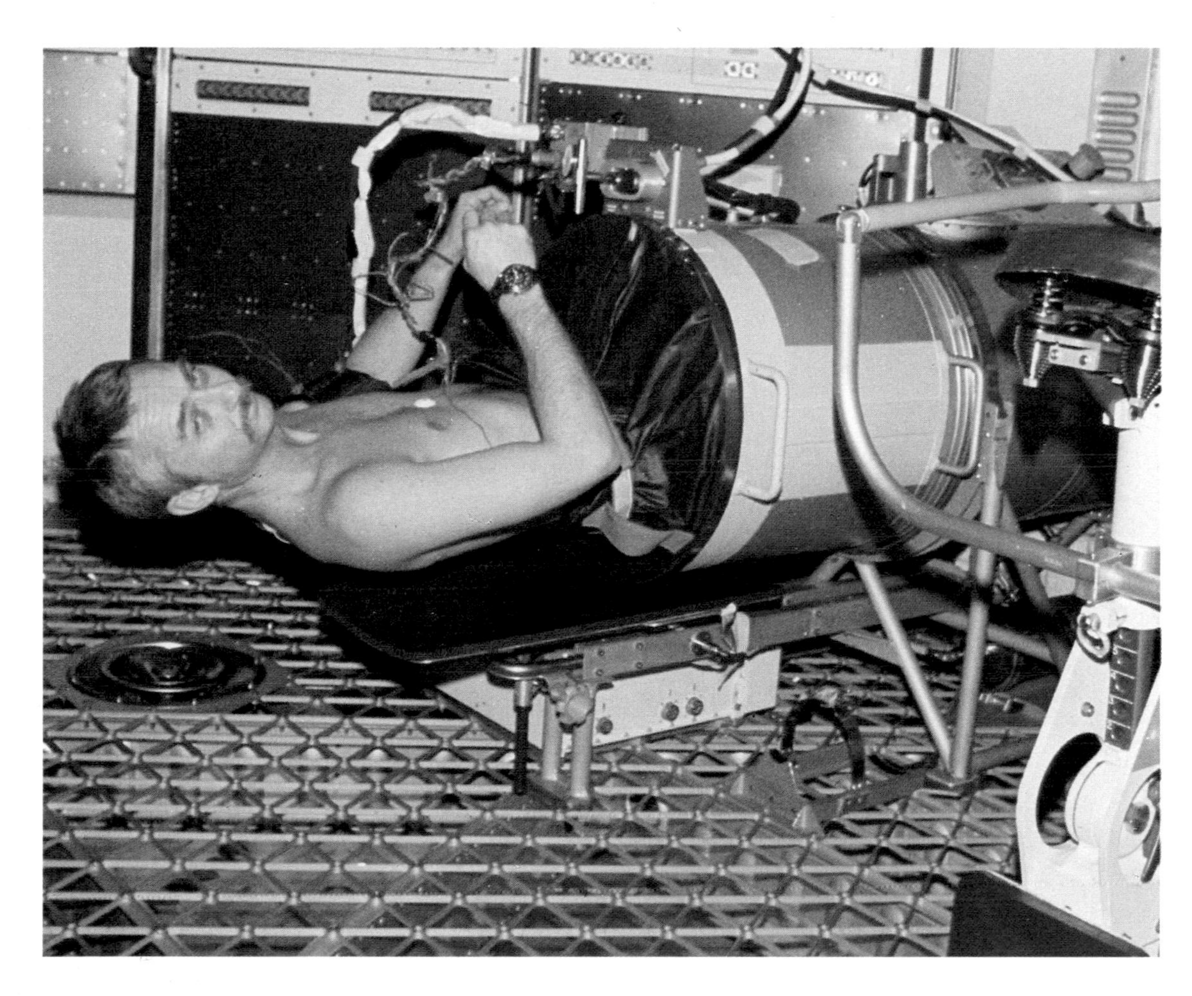

▲ **Perfect balance**
Balancing tricks are child's play in zero-g. Skylab 4 astronaut Edward Gibson shows how to hold up Gerald Carr on only one finger. Who is he kidding?

The Skylab cluster

The layout of Skylab is illustrated on page 189. With the Apollo spacecraft docked with it, Skylab measured 36 meters (119 feet) long. It had an Earth weight of about 90 tonnes. The main part was the orbital workshop (OWS), which was the part based upon the S-IVB. It housed the main living quarters, comprising wardroom, sleeping compartment, waste-management compartment (toilet facilities) and experiment compartment, in which the astronauts conducted most of their experiments – the main control center was also located there. These living quarters formed the lower level of a two-storey structure in the former hydrogen tank of the S-IVB, which was some 15 meters (50 feet) high and 6 meters (20 feet) in diameter. The smaller oxygen tank underneath the living area floor was used to store trash.

The forward compartment above the living area was ringed with storage lockers and provided room for the long-stay crews to let off steam with spectacular running and gymnastic displays. An airlock module (AM) led from this into the multiple docking adapter (MDA). The MDA was equipped with docking ports for the Apollo spacecraft that ferried up the astronauts. On the MDA was mounted the Apollo telescope mount (ATM), which carried a package of instruments for observing the Sun. The control center for the ATM was located in the MDA. Electrical power was provided by solar panels on the ATM and the OWS.

Three by three

The planned launch on 15 May 1973 of the first manned mission to Skylab (designated Skylab 2) did not materialize. The apparently perfect launch of Skylab itself the previous day was anything but. Premature deployment of the meteoroid/heat shield during launch had left a gaping hole in Skylab's side, ripped off one solar panel and jammed another. The crew, Charles Conrad, Joseph Kerwin and Paul Weitz, eventually lifted off 10 days late and managed to save the mission by DIY repairs while spacewalking (see page 154). This emphasized one of the great advantages of sending people rather than robots to explore the space frontier.

The Skylab crew settled down to a routine that included a heavy workload. The experiments were centered on three main areas – medical, solar observations and Earth resources. In addition the crew carried out experiments in materials science, melting and crystallizing different materials in an electronic furnace.

In the medical experiments the astronauts regularly took blood samples from one another and checked the blood circulation, or cardiovascular system. They took photographs of the Sun at various wavelengths, and monitored and photographed the Earth through a package of instruments called the Earth resources experiments package (EREP). By mission's end they had acquired nearly 30,000 pictures of the Sun and nearly 16 km (10 miles) of geophysical data.

◀ **Fond farewell**
The Skylab 4 crew make a final pass over the space station before they return to Earth on 8 February after 84 days in orbit. Prominent in this view is the jury-rigged parasol over the damaged exterior, which proved remarkably effective as a heat shield.

▼ **Fiery fountain**
Skylab's solar telescopes capture this image of one of the biggest prominences ever seen on the Sun. These fiery fountains of incandescent gas loop hundreds of thousands of kilometers through the outer solar atmosphere, traveling along magnetic force lines.

◀ **Weightless web**
Space spider Arabella makes an excellent attempt at web-spinning on Skylab's second manned mission. She survives to return to Earth, though her sister Anita dies on the job.

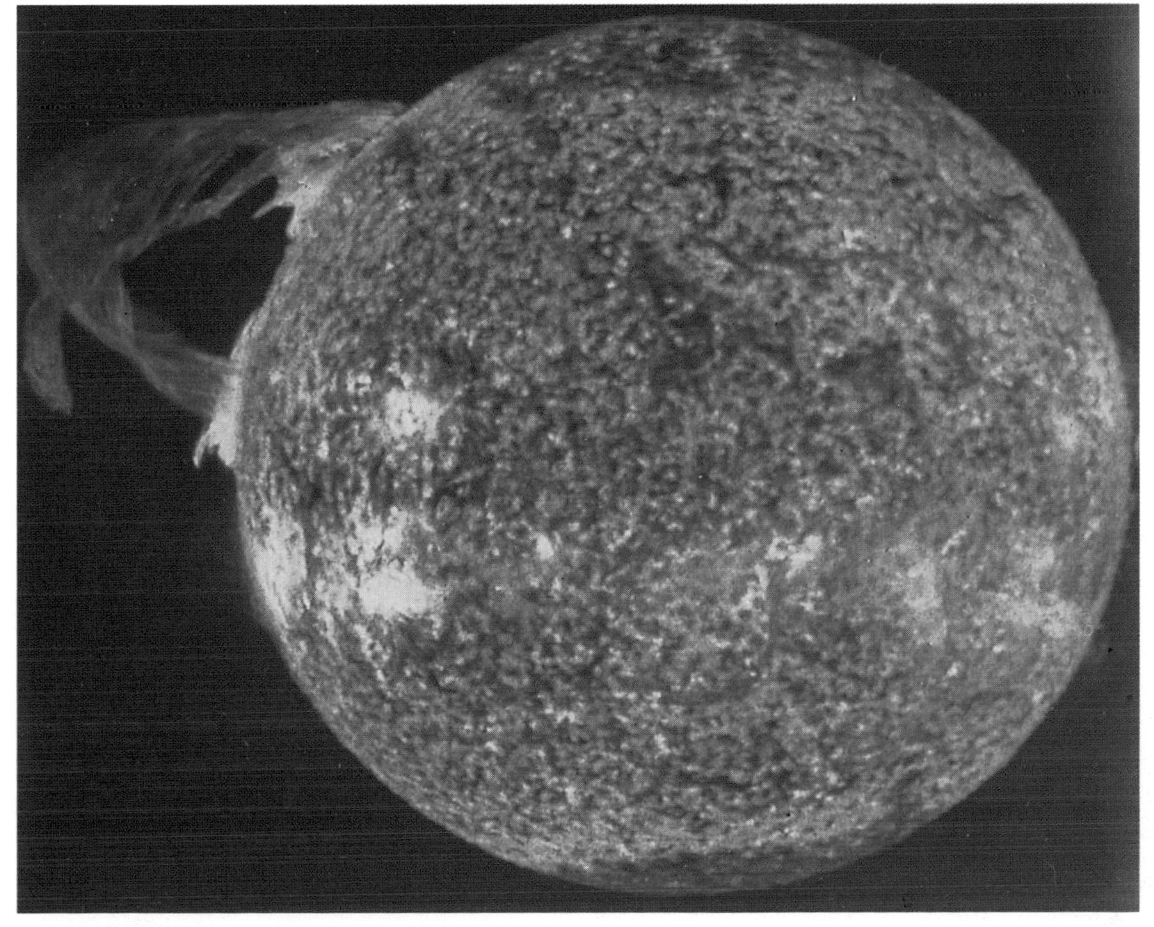

The astronauts ate their food in a civilized manner on a tray on the wardroom table and with conventional cutlery. They were permanently hungry, even though the food gave them the calories they required. They reckoned that they must have left their taste buds on the ground because they found the food flavorless. They recommended that spicier foods should be included in subsequent missions.

To prevent their muscles weakening too much in the zero-gravity environment, the astronauts had daily exercise sessions on a bicycle ergonometer. Conrad, often pedalling with his hands instead of his feet, exercised most and predictably he ended up fitter than the others.

When the crew splashed down on 22 June, the crew had established a new space endurance record of 28 days, during which time they had traveled 395 times around the Earth – a distance of nearly 14 million miles. As the recovery ship raced to pick them up Conrad radioed: 'Everybody here is in super shape.' And they were.

Doctors who checked them out over the next weeks reported the typical weakening of the heart; weakening of the bones due to loss of calcium; and reduction in size and flabbiness of the calf muscles due to atrophy – this is generally termed the 'birds' legs' effect. But after a few days the crew had more or less fully recovered their strength. Conrad commented significantly at a press conference: 'I'd say very definitely that the average man or woman could fly in space.' That was a prophecy, which is now coming true in the shuttle era.

From strength to strength

Space sickness – or as it is properly called space adaptation syndrome – did not trouble the Skylab 2 crew even in the first few days in space. The next crew to visit Skylab, Alan Bean, Owen Garriott and Jack Lousma, were not so lucky. They began to feel spacesick as soon as they got on-board Skylab on 28 July 1973. Fortunately, their bodies soon adapted to zero-g and the sickness passed.

Among the urgent tasks they had to carry out during their planned 59-day mission were the installation of a new sunshade over the damaged workshop and the fitting of new gyroscopes to the space station's attitude-control system. This they accomplished during taxing but otherwise uneventful spacewalks. So efficiently did the astronauts get through their workload that on the fourth week they asked for more! And they also recommended that more be included in the next mission.

Among the investigations they carried out was one suggested by a high-school student: can spiders spin webs in zero-g? Accordingly they took two spiders along with them into space to see how they managed. The spiders, called Anita and Arabella, both spun quite well despite their strange environment.

A major problem developed during the mission, not with Skylab, but with the Apollo ferry. Its thrusters began leaking, prompting fears that it might lose vital steering functions that would be necessary when the crew returned. Plans were made at the Cape for a rescue mission, which would be piloted by two of the astronauts scheduled to make the final visit to Skylab later in the year. In the event, no rescue was needed. The Apollo spacecraft performed as it should and took the crew back to Earth on 25 September. After nearly two months, 892 orbits and 38 million km (24 million miles), they were pronounced fitter than the first Skylab crew!

The final Skylab visit (Skylab 4), by Gerald Carr, Edward Gibson and William Pogue, began on 16 November. It was delayed so that the crew could observe Comet Kohoutek, dubbed the 'comet of the century'. Unlike the previous crew, the astronauts soon began grumbling about the heavy workload and began making mistakes. Friction and irritability persisted for nearly six weeks, when they had their first real day off. Then after a heart-to-heart chat with ground control, the atmosphere began to improve. Explained Pogue: 'When I tried to operate like a machine, I was a gross failure.... We've got to appreciate a human being for what he is.'

The remainder of their mission was comparatively harmonious. They splashed down on 4 February 1974, 84 days, 1260 orbits and 55 million km (35 million miles) after launch. The $2.5 billion Skylab project was over, after 513 man-days in space. Not for nearly 10 years would another US space laboratory take to the skies. Then it would be as a cooperative project with the Europeans.

Death throes

After the last crew's triumphant return from Skylab, the problem was what to do with it now. Because of its size, NASA knew that it could only remain in orbit for a few years. There are still tenuous traces of atmosphere even at Skylab's altitude (435 km, 270 miles). And they exerted drag on the space station, slowing it down and eventually causing it to plummet back to Earth. Because it was so big, quite large chunks could survive the re-entry through the atmosphere and threaten life and limb on Earth. NASA's original plan was to wait until the space shuttle became available and then use it to boost Skylab into a much higher orbit, where maybe it could be used again.

But delay in the launch of the space shuttle ruled this out. Also Skylab's orbital decay was accelerated by increased solar activity. By early 1979 it was clear that Skylab had only a matter of months to live. The end finally came on 11 July, when the space station began to break up over Ascension Island in the Atlantic on its 34,981st orbit. The fiery debris began to rain down in a 160-km (100-mile) swathe, or footprint, that took it ultimately over Western Australia.

In Perth, residents saw luminous bodies flash overhead. Said one: 'It was like Tinker Bell waving a magic wand, sparks whirling everywhere.' Fortunately, most of the debris came down on the sparsely populated Nullarbor Plain. Cassandras the world over, who had predicted death and destruction, were proved wrong. But it all could have ended differently.

Laboratory in space

After Skylab, America concentrated its main energies in the space field into getting the space shuttle off the ground. Eventually, in April 1981 it did, two years later than planned. The main purpose of the shuttle was envisaged as a space truck, able to launch satellites into orbit, two or three at a time, and at a lower cost than conventional expendable rockets.

Although the space shuttle orbiter was more spacious than the early spacecraft, there was still little room on-board for scientific

▶ **Science in orbit**
Up in space Spacelab remains in the shuttle orbiter's payload bay. The scientists conduct their research, round the clock, in the pressurized laboratory module. An access tunnel connects the laboratory with the mid-deck of the orbiter, where the crew live off-duty. The space station is pictured in its long-module configuration, with two pallets. One of the crew is spacewalking to check the pallet instruments.

▶ (far) **ESA in space**
With emblems of the European Space Agency (ESA) well to the fore, Spacelab is pictured on its space debut in the open payload bay of *Columbia* in November 1983. This view, taken from the aft crew station on the upper deck, shows the access tunnel to the laboratory.

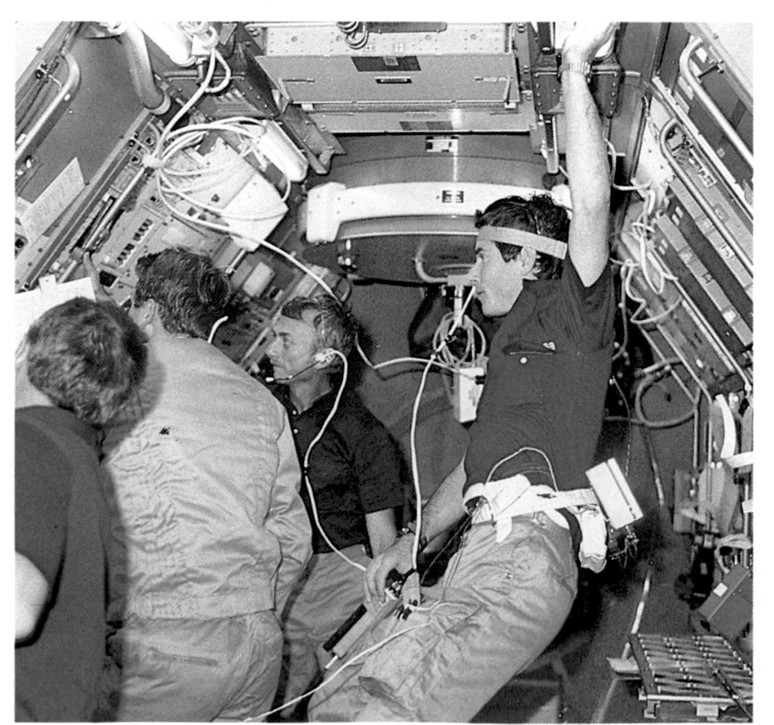

◀ **Scientists at work**
Intent on their labors, four of the six crew of Spacelab study results displayed on a video screen. At right is West Germany's Ulf Merbold, the first non-American to fly with a US crew; and at the back is that veteran of Skylab 3, Owen Garriott.

◀ (far) **Checkout**
All but ready for flight, Spacelab 1 is put through final checks in the Operations and Checkout Building at the Kennedy Space Center. Later it will be installed in *Columbia*'s payload bay for its November 1983 launch.

experimentation. So, would US space science have to wait until the 1990s when the NASA space station became operational? This would be unthinkable, with the Soviets every year expanding increasingly into space in their Salyut space stations (see page 198). So it was with this in mind that NASA, in collaboration with the European Space Agency, ESA, established the Spacelab project.

Spacelab is a fully equipped scientific laboratory designed to fit into the cavernous payload bay of the shuttle orbiter. It remains there all the time. It is made up of a number of modules and platforms, which can be linked together in a variety of configurations. The first Spacelab flight took place on 28 November 1983. It was carried into space by orbiter *Columbia* on its sixth, and the shuttle's ninth mission (STS-9).

Record breaker

The flight set a number of records. It was the first time six astronauts had been carried into space at the same time. It marked the sixth flight by its commander, John Young. It was the first flight to carry two non-professional astronauts, the so-called payload specialists. They were Byron Lichtenberg and, from Germany, Ulf Merbold. Merbold became the first non-American to fly on a US spacecraft. The remaining members of the crew were Robert Parker and Skylab veteran Owen Garriott.

Spacelab 1, as the flight was designated, had the so-called long module configuration, which featured a double-segment pressurized laboratory. This measured about 7 meters (23 feet) long and 4 meters (13 feet) across. Attached to it was a U-shaped platform, or pallet, which carried instruments and was open to the space environment.

Working round the clock in three-man shifts, the crew enthusiastically tackled over 70 experiments in the life sciences, astronomy and solar physics, Earth observation and materials technology. They processed some data on-board with their computers, but much was transmitted back to Earth scientists via the recently deployed TDRS (tracking and data relay satellite). There were teething problems with these communications and some data was lost, but in the main the flight went well – so well that NASA extended the planned 9-day mission to 10.

The only major hiccup occurred not in Spacelab but in *Columbia*, when two of its flight computers failed one after the other during re-entry maneuvers. Fortunately, there were five on the orbiter, and *Columbia* touched down safely.

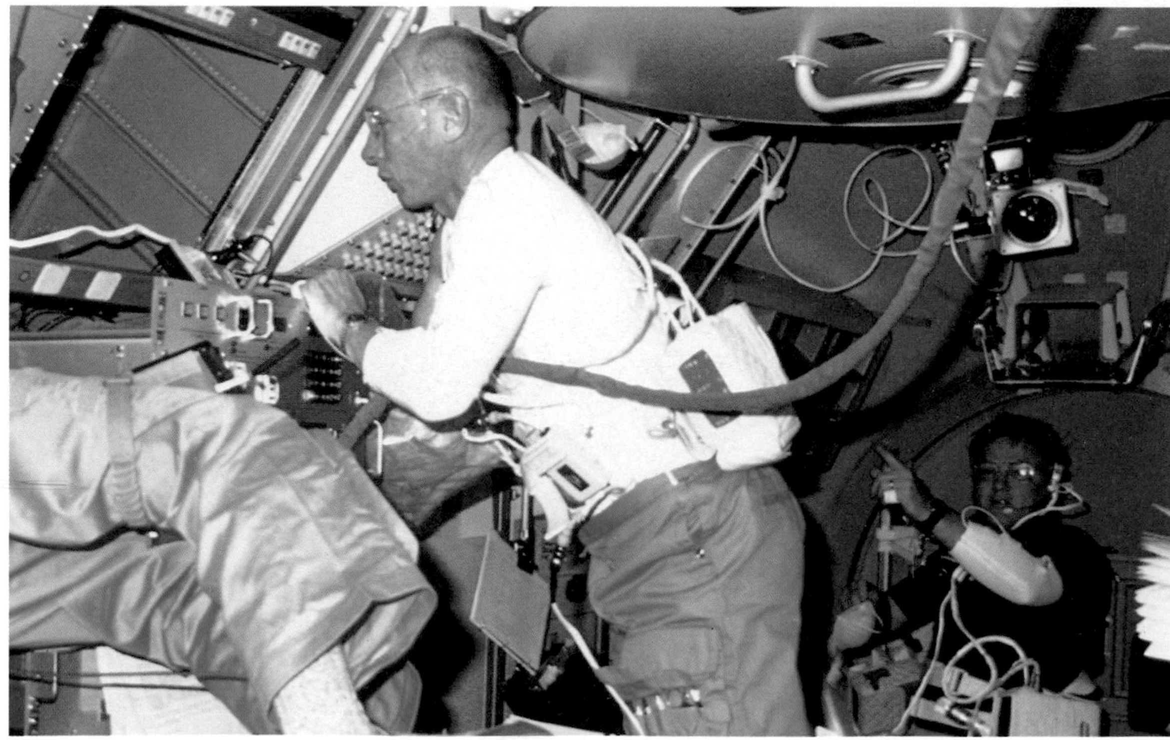

▼ **Fixing it**
On the Spacelab 3 mission Taylor Wang disappears into the innards of the space laboratory to conduct repairs. Looking on are William Thornton (center) and Donald Lind.

▼ (right) **Peek-a-boo**
Also on Spacelab 3, one of the two squirrel monkeys in residence is studying the effects of weightlessness on William Thornton.

◀ Blood-letting
West German payload specialist Reinhard Furrer bares his arm ready for the needle during one of the regular blood-sampling exercises during the Spacelab D1 mission in October 1985.

Monkeys, rats and flies

Spacelab's second mission (actually designated Spacelab 3) began on 29 April 1985. It was carried into space by *Challenger* on mission 51-B. The seven-man crew included payload specialist Lodevijk van den Berg from the Netherlands. He was one of the three men on-board who were over 50 years old – William Thornton was the oldest at 56, proving that advancing age is no longer a barrier to space flight.

This Spacelab flight, also in the long-module configuration, also marked the first time that men and animals had flown into space together. Two squirrel monkeys and 24 rats were on-board. Preflight reservations about carrying a mini-zoo were borne out when pilot Robert Overmyer found monkey feces floating past him in the cockpit! The crew conducted research into crystal growth and fluid mechanics. They were also fortunate to witness a magnificent display of the aurora – the shimmering ever-moving curtains of light that occur in polar regions.

The Spacelab 2 mission (51-F), which lifted off on 29 July 1985, began inauspiciously with an 'abort to orbit', when one of *Challenger*'s main engines shut down due to apparent overheating. It was purely an instrumental mission devoted mainly to astronomical research. The various instruments and telescopes were mounted on pallets, with an 'igloo' control center.

Astronaut Willy

The final Spacelab flight in 1985 was a West-German dedicated mission, Spacelab D-1. It went up in *Challenger* (mission 61-A) on 30 October 1985. It was fated to be *Challenger*'s last flight into orbit. Spacelab was once again in its long-module configuration. One of the most interesting of its experiments involved the use of the space sled. This comprised a seat on which an astronaut was accelerated along a 7-meter (23-foot) long track. He wore elaborate headgear to expose his ears and eyes to different stimuli. The experiment was devised to gain greater insight into the problems of space sickness.

The biological experiments included investigations of the growth of corn and cress seeds, frog spawn and fruit flies. One of the fruit flies escaped from its container and began to fly around the cabin. Named Willy by the Spacelab crew, it was later discovered dead in a filter. Later, a spokesman from Spacelab's scientific control center at Oberpfaffenhofen in Bavaria handed the press an obituary for 'our bold little astronaut'!

▼ Triple link-up
The size of the Salyut space station complex is well illustrated by this exhibit in the Cosmos Pavilion at the Exhibition of USSR Economic Achievements in Moscow. It shows a Soyuz manned spacecraft (left) and an unmanned Progress ferry (right) docked at opposite ends of Salyut 6.

▶ **A year later**
Three cosmonauts give a Press conference after touching down in Kazakhstan on 21 December 1988. Jean-Loup Chretien (left) has returned after nearly a month in the *Mir* space station. The other two, Vladimir Titov (middle) and Musakhi ('Musa') Manarov, have not felt the Earth under their feet for a year!

▼ **A happy ship**
Pictured on-board Salyut 6 in September 1980 are, from the left: Valery Ryumin, Arnaldo Mendez, Yuri Romanenko and Leonid Popov. Mendez, an ex- shoeshine boy, is a cosmonaut 'guest' from Cuba. Ryumin and Popov go on to break the space endurance record with a stay of nearly 185 days in orbit.

Success with Salyut

While Western space science has tended to be practiced rather in fits and starts, Soviet space scientists have been more methodical, building up a formidable bank of knowledge through their continuous Salyut space-station program and more recently with *Mir*. After the initial failure of Salyuts 1 and 2, exacerbated by the death of the Soyuz crew (see page 188), they made gradual progress with Salyuts 3-5. But it was not until the launch of Salyut 6 in September 1977 that the real record-breaking began.

Salyut 6 was of much the same design as its predecessors, being made up basically of three cylinders of different diameters, the biggest about 4 meters (13 feet) across. Its length was about 14.5 meters (47 feet). Unlike the earlier Salyuts it had two docking ports fore and aft which allowed two craft to dock at the same time.

The purpose of this double port soon became evident. The first crew to occupy Salyut 6 went up in Soyuz 26 on 10 December. On 11 January 1978, Soyuz 27 also docked with the space station. It was the first triple link-up in space and looked forward to the type of multiple spacecraft complex that the Soviets envisaged.

After the visitors returned to Earth in Soyuz 26, came another breakthrough. On 20 January a remote-controlled and unmanned space ferry called Progress docked at the free port. It was the success the Soviets needed to push ahead with its long-stay cosmonaut program. Using Progress ferries solved the problem of

▶ ***Mir* in orbit**
Though broadly similar to its Salyut ancestors, *Mir* differs by having a multiple docking module at one end (the lower end in this picture). Spacecraft can also dock at a port at the other end.

keeping the cosmonauts supplied with food, post and propellants, as well as any other equipment the cosmonauts might need in an emergency.

Counting the days

The visit by Progress enabled the Soyuz 26 cosmonauts, Yuri Romanenko and Georgi Grechko, to stay in orbit for 96 days, beating the US Skylab record of 84 days. On 15 June Soyuz 29 went up to Salyut and its crew, Vladimir Kovalyonok and Alexander Ivanchenkov, did not return until 2 November, 140 days later. During their marathon mission, the cosmonauts had been visited twice by other crews and been re-supplied by three Progress ferries.

And so the impressive numbers game continued in 1979. On 25 February, Vladimir Lyakhov and Valery Ryumin went up in Soyuz 32 and did not return for 175 days, on 19 August. Incredibly, Ryumin was back in space for another marathon the following spring (9 April). This time he and his crewmate Leonid Popov spent 185 days in orbit. Then came the launch of Salyut 7, which by the end of 1984 had accommodated crews for 211 days (Soyuz T-5) and 237 days (Soyuz T-10).

The jinx ship

Despite the marathon flights in Salyut 7, it eventually acquired the reputation of being a jinx ship. The crew who went up in Soyuz T-9 in June 1983, Vladimir Lyakhov and Alexander Alexandrov, in the second week of September almost had to make an emergency evacuation of Salyut following a fuel leak, but this eventually proved less serious than supposed. Soon afterwards one of the station's three solar panels failed, starving it of electricity. To remedy the situation, two more cosmonauts on 27 September prepared to go up to fit new solar panels. But a few seconds before launch, the launch rocket exploded. Miraculously, the Soyuz's escape rocket system worked perfectly and lifted the crew out of the ensuing fireball. This was the first time in space history that there had been such an on-the-pad abort.

The Salyut 7 jinx stayed dormant in 1984, but in January 1985 came back. The station began drifting out of control and its water-supply system sprang a leak, causing flooding and electrical shorts. In March there came rumors that the space station was to be abandoned, but in fact plans were being made for a repair job. And three months later two cosmonauts in Soyuz T-13 went up and managed to make Salyut 7 habitable again against all the odds.

The final chapter in the Salyut 7 saga occurred on 11 November when the three cosmonauts on-board were suddenly recalled 'because of illness'. What apparently happened was that the very young commander, Vladimir Vasyutin, cracked under the stress of his first space-flight command. His colleagues said he became 'a bag of nerves'.

◀ **Space spiderman**
In November 1985 Jerry Ross demonstrates on shuttle mission 61-B the techniques that astronaut construction engineers might use to build the truss structures for the space station *Freedom*. Using the method known as ACCESS, he makes the task look deceptively simple.

Mir takes over

Two of the cosmonauts involved on the 237-day record-breaking trip in Salyut 7, Leonid Kizim and Vladimir Solovyov, were first to enter its successor, *Mir* (meaning 'Peace'), launched on 19 February 1986. They went inside on 15 March. On 6 May they cruised over to Salyut 7 for a final visit in the first transfer between space stations in orbit. They returned to *Mir* on 26 June. On 2 July Kizim set another world record by becoming the first person to clock up a total of one year in space. It was an auspicious start for the new spacecraft.

Space station *Mir* looks much like Salyut outside, except that it has an extra module at the nose. This is a multiple docking module with no less than six docking ports. *Mir* has longer solar panels than Salyut and looks, according to Kizim, 'like a white-winged seagull soaring above the world'. The major differences, however, lie inside.

Mir is outfitted primarily as a habitation module, with relatively comfortable cabin accommodation for a crew of up to six. Experiments take place in add-on science modules, like the one called Kvant ('Quantum'), which docked with it at the end of March 1987. As time goes by it is expected that *Mir* will grow with the addition of several new modules at the vacant docking ports.

In *Mir* the space endurance records have continued to tumble. On 29 December 1987 Yuri Romanenko was among the three-man crew of the returning Soyuz TM-3. He had spent 326 days in orbit. He was in incredibly fine shape. He could stand up, albeit shakily, shortly after he touched down, and the next day managed a 100-meter jog!

When Romanenko left *Mir*, two other cosmonauts, Musa Manarov and Vladimir Titov, had already been in space for six days. They remained there for another 360 and became the first human beings to spend a year continuously in space. According to some experts, this will give the Soviets sufficient information to be able to forecast the likely effects on astronauts of a manned mission to Mars, a well-known Soviet goal.

Freedom in space

By the mid-1990s *Mir* may well have expanded into an extensive multimodular complex. Indeed there may well be other *Mir*-type space stations in orbit. Orbital activity will become even more intense as the first launches take place of the hardware that will be assembled into the first permanent Western space station. NASA will be the prime mover behind the station, but it will also feature significant contributions from Canada, Europe and Japan. So it will truly be an international station.

The go-ahead for the space station was given early in 1984. In his State of the Union message on 5 January 1984 President Reagan said: 'We can follow our dreams to distant stars, living and working in space for peaceful, economic and scientific gain. Tonight I am

▼ **International *Freedom***
This close-up of the core complex of *Freedom* shows the four main modules in which the crew will live and work. Cutaway is the US laboratory module. Behind is the habitation module, also built by the US. The laboratory modules at the left are provided by the European Space Agency and Japan's National Space Development Agency (NASDA)

directing NASA to develop a permanently manned space station and to do it within a decade.' In 1988 the President named the space station *Freedom*.

Unlike Skylab and Salyut, *Freedom* will not be launched bodily into space on a launch vehicle – it is too big for that. It will be launched piece by piece and then put together in orbit. The space shuttle will be the main means of lifting the hardware into orbit, beginning in 1994.

Baseline configuration

The baseline design of *Freedom* has now been finalized. It has a backbone, or keel, some 150 meters (500 feet) long. This is a trussed structure that may well be put together in a similar manner to the EASE/ACCESS beams built by the astronauts on shuttle mission 61-B. The major part of the station is made up of four pressurized modules, which are attached to the middle of the keel. They are interconnected by so-called resource nodes. The modules are about 15 meters (50 feet) long and 4.5 meters (15 feet) across. At this size, they can be accommodated in the shuttle orbiter's payload bay.

NASA is providing two of the modules – one is the habitation module in which the crew, probably numbering six, will live; the other is a laboratory module. The other modules are supplied by Japan and ESA. Japan's design is called JEM (Japanese experiment module). It consists of a pressurized module, a logistics module to support experiments, a robot manipulator and an open platform for exposing instruments and samples to space.

Europe's space-station contribution is organized under the so-called Columbus program. This includes a laboratory unit, the APM (attached pressurized module); a free-flyer laboratory called the MTFF (man-tended free flyer); and a polar platform (PPF). In the construction of the APM, ESA will draw heavily from its experience with Spacelab. The MTFF consists of a pressurized module, which astronaut-scientists can visit, together with an unpressurized module. It is anticipated that eventually the European space plane Hermes will ferry European astronauts to the MTFF for routine servicing. Hermes is a delta-winged craft designed to be launched by the yet-to-be developed Ariane 5 rocket.

Other major elements of *Freedom* include the huge solar panels located at each end of the keel, designed to supply 75 kilowatts of electrical power. Canada will provide a mobile servicing system, a 'roving space robot which will be the arms and hands' of the space station. It is designed using technology borrowed from the highly successful robot manipulator arm used on the shuttle orbiter. It will be involved in unloading shuttles, capturing and relaunching satellites and supporting astronauts working in space. As its name implies, it will be capable of being moved to where it is needed.

Another useful piece of hardware that will be employed at the space station will be the orbital maneuvering vehicle (OMV). The OMV is a space tug, which will be used, for example, to capture distant satellites and take them back to the space station for servicing. It is essentially a remote-controlled reusable rocket vehicle, equipped to dock with satellites and other hardware that need to be shifted around in orbit.

▲ Service call
At the turn of the century a shuttle orbiter closes in to dock with the now complete space station, orbiting at an altitude of about 400 km (250 miles). It is bringing up a crew for the designated 90-day tour of duty. Longer stays of up to a year may be permitted for space medicine research but will be infrequent.

◀ Weightless bliss
The luxuries of life are not to be denied to *Freedom*'s residents. The design of the whole body shower is already well advanced. Here a plexiglass prototype is being tested in weightless flight in the zero-g KC-135 aircraft. The test subject is rinsing her hair after a shampoo and shower. She is using a water spray to rinse, and a vacuum hose to suck up excess water.

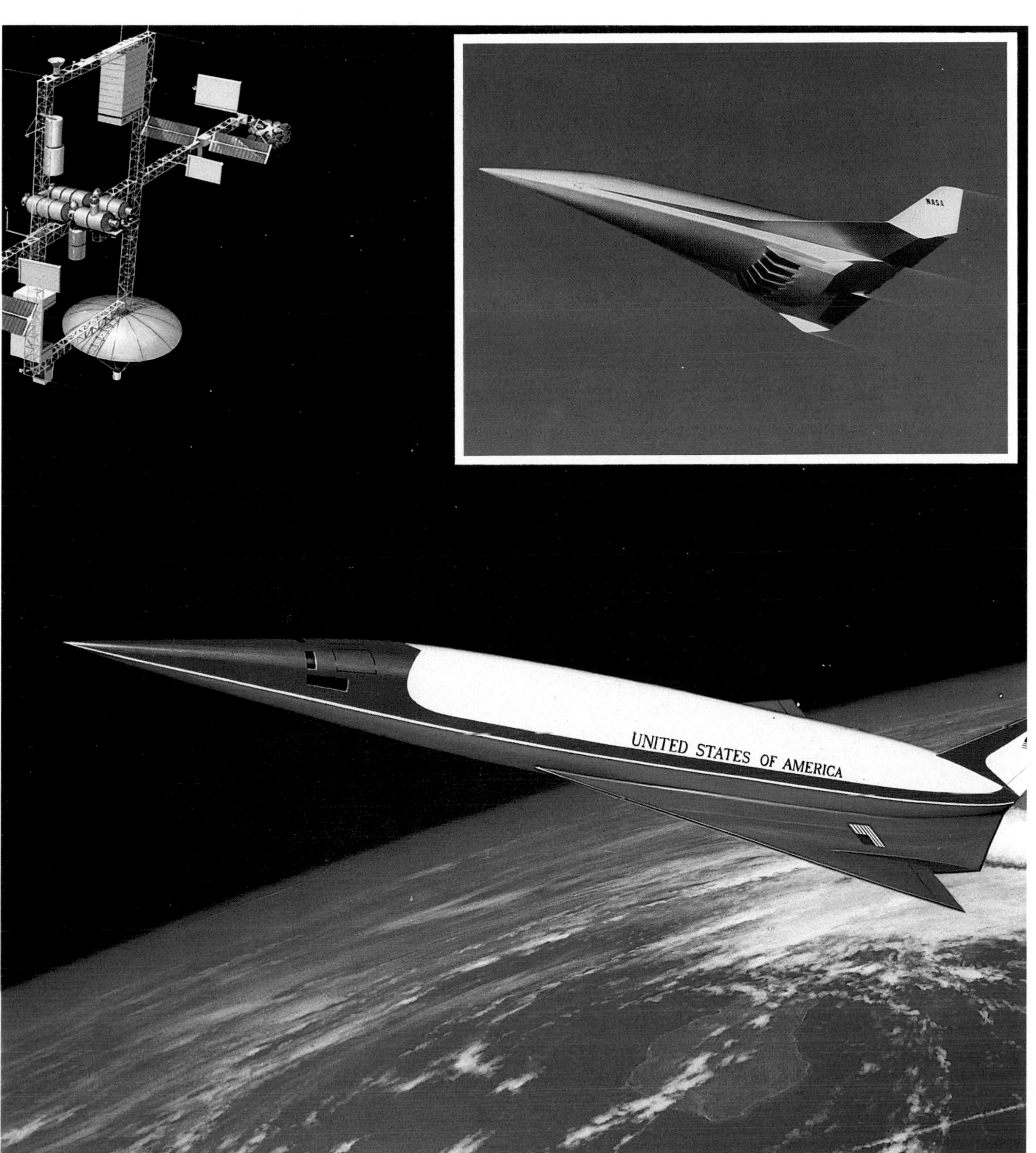

Orient Express
Accelerating towards the space station early next century is the US space plane, which is capable of a conventional runway take-off yet is able to reach obital speeds. Because airliner versions could speed from Washington to Tokyo in under two hours the plane has been nicknamed the Orient Express. More correctly it is termed the National Aerospace Plane, designated the X-30. Europe has a rival plane under consideration called Hotol (horizontal take-off and landing). Both feature a revolutionary kind of engine and extract oxygen from the atmosphere for use in space. Inset is another design for the US space plane.

Shuttle HLLV

It is reckoned that upwards of 20 shuttle launches will be needed to support the first phase of space-station construction, which is quite a daunting prospect. However, an alternative method is being looked into, which would halve the number of launches and also shorten construction time. This involves the use of a shuttle derivative called the Shuttle C.

The Shuttle C is a proposed unmanned heavy cargo version of the shuttle. It uses features of the existing shuttle, but replaces the reusable orbiter with an expendable payload carrier, or cargo element. It has a much greater payload capacity than the shuttle – up to 80 tonnes. The new vehicle uses ex-shuttle main engines, fed by fuel from an external tank and solid rocket boosters at lift-off. So it dovetails in with existing shuttle facilities in assembly and on the launch pad. In a typical mission to *Freedom*, Shuttle C would deliver the cargo element into orbit, where it would be parked until it was needed. An OMV would then dock with it and guide it to the space-station site. After its payload has been extracted, the cargo element's engines would be fired in a de-orbit burn. It would then descend and break up in the atmosphere during re-entry.

Distant horizons

The space station will provide a permanent base for prolonged scientific investigations of the kind which took place on Skylab; which takes place fitfully on Spacelab; and which is practiced more continuously on *Mir*. The main areas will be space medicine, the life sciences, materials technology, Earth resources, astrophysics and astronomy generally. Early spin-offs from space-station investigations are expected to be the development of industrial-size manufacturing facilities for the preparation of ultra-pure drugs and vaccines, and the production of superalloys and superfine semiconductor crystals for electronics.

The development of construction skills required in the assembly of the space station could well lead in the early decades of the 21st century to more ambitious in-orbit construction projects, maybe to such exotica as satellite solar power stations. By then the station, or an offshoot from it, will be functioning as a spaceport for handling and servicing shuttle vehicles traveling to the Moon and even to Mars.

It is envisaged that a return to the Moon could take place as early as the year 2000. This time the lunarnauts will establish an outpost that will evolve into a permanently manned base. Initially the base will have to depend on regular supplies from Earth. But eventually it will be able to develop mining and manufacturing facilities to extract and process materials of construction from the surface. Thanks to the Apollo explorations, we know that such activities will be feasible. Even the oxygen for life-support could be extracted from the rocks.

Mission to Mars

While erstwhile colonists are establishing a base on the Moon, others may be heading farther afield to the most enigmatic of our heavenly neighbors, the Red Planet Mars. Mars has always fascinated man because of its fiery color. Many eminent people, including the American astronomer Percival Lowell, were convinced that there was life on Mars. Close-up surveys of the planet by space probes, however, show that the chances are remote.

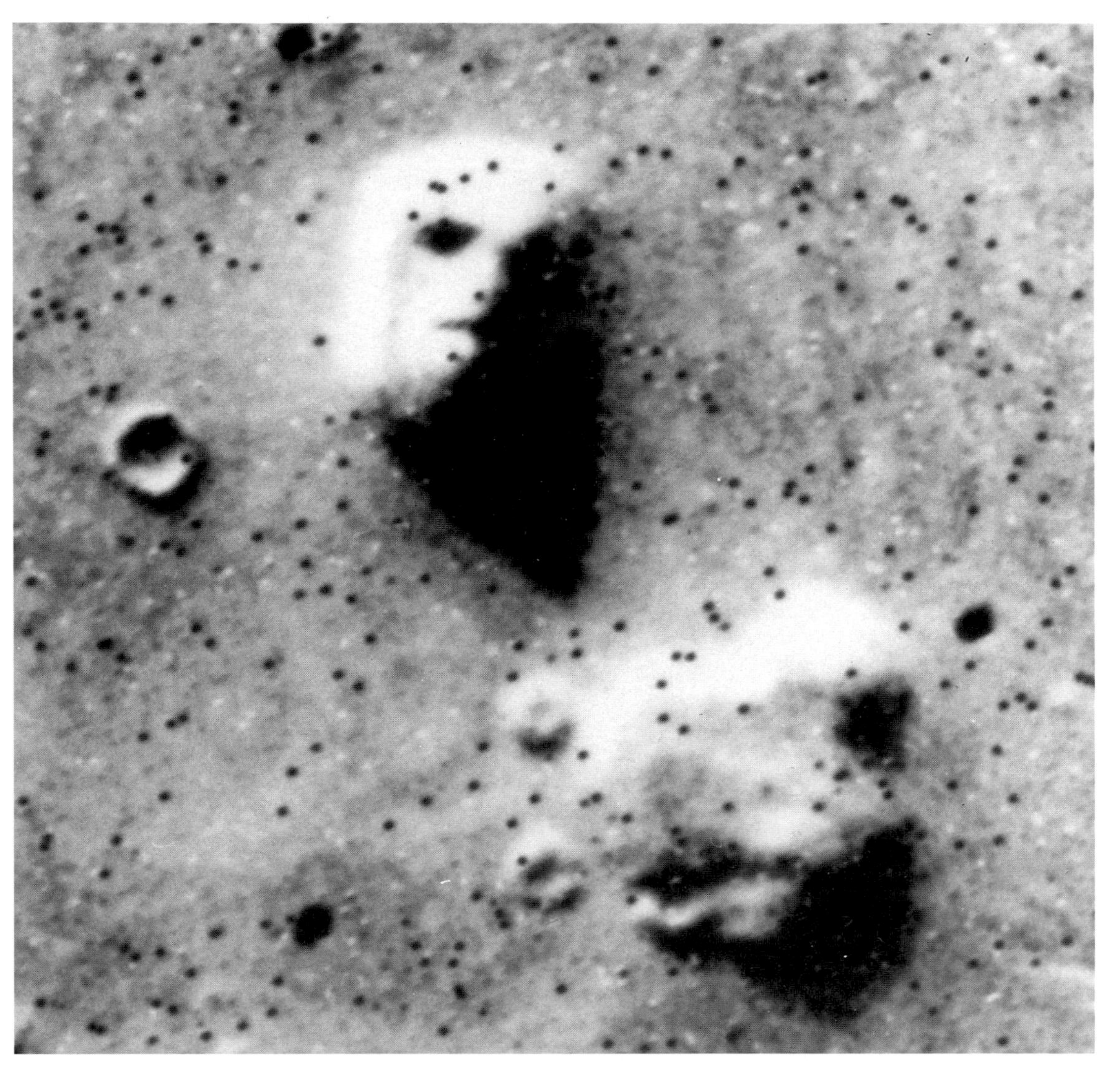

▲ A Martian face
In 1976 an orbiting Viking space probe photographs a feature on Mars that bears an uncanny resemblance to a face. Is it natural or can it be the handiwork of a long-since vanished Martian civilization? It is known that in the distant past, Mars had a much milder climate and ran with surface water. So is it possible that an intelligent Martian race evolved?

◀ Lunar base
When man returns to the Moon early next century, he will set up a permanent base. In a decade or so the base will start to become self-sufficient by extracting materials from the lunar surface. This illustration shows an industrial complex set up to extract oxygen.

Martian base
Mars will be the next stop on man's journey of exploration into the solar system. A base like this could grow up by the middle of next century. The Martian explorers will benefit from the experience gained by the setting-up of the Moon base and establish similar facilities.

▲ **Mars rover**
Before man goes to Mars, robot explorers will be sent to take a closer look. One concept envisages that a rover (above) would collect and analyze soil samples. It would also transfer some samples to a rocket. The rocket will carry them up to an orbiter. which would return them to Earth.

Nevertheless, Mars will be the target of the first manned interplanetary mission because it is the only planet in the solar system apart from the Earth that a human being could set foot on and live to tell the tale. It is cold, but not too cold – noonday temperatures at the equator can rise above freezing.

A round trip to Mars would, to put it mildly, be a very hazardous undertaking. It would take up to three years to travel back and forth across the great divide that separates us, a distance of at least 55 million km (35 million miles). Cosmonauts have shown that there appears physically to be no limit to the time human beings can stay in space, providing regular exercise is taken. The other problems of such a trip are daunting, however. Once the astronauts set out on such a trip, they would be quite beyond rescue. As yet no life-support system exists that could sustain a crew for three years, but experiments in recycling materials have proved encouraging. Also, no suitable rocket exists that would have the capability of such a journey. But the answer could lie in nuclear-powered ion engines. This would require the crew to be shielded from the nuclear radiation. They would have to be shielded in any case from the effects of cosmic radiation, the penetrating and potentially lethal rays that emanate from the Sun and the cosmos generally.

So will men go to Mars, given such apparently insurmountable problems? The answer has got to be, Yes. Mars is another cosmic Everest to be climbed, maybe simply because it's there. And the chances are that it will be climbed sooner rather than later. After Mars, who knows? Some space gurus have suggested the asteroids, which could be mined for minerals. Some have put forward exotic schemes to turn hellish Venus into a balmy twin of Earth. Others have designed spaceships to travel to the stars.

This is the stuff of science fiction. But science fiction does have a habit of becoming science fact. And that doyen of science-fiction writers Arthur C. Clarke has observed: 'There is no way back into the past; the choice, as [H.G.] Wells once said, is the universe – or nothing.'

USA
NASA
Challenger

THE MISSIONS

1961

Mission 1 – Vostok 1 – 12 April
Crew: Yuri A. Gagarin
Duration: 1 hr 48 min

Mission 2 – *Freedom 7* – 5 May
Crew: Alan B. Shepard
Duration: 15 min 28 sec

Mission 3 – *Liberty Bell 7* – 21 July
Crew: Virgil I. Grissom
Duration: 15 min 37 sec

Mission 4 – Vostok 2 – 6 August
Crew: Gherman S. Titov
Duration: 1 day 1.3 hr

1962

Mission 5 – *Friendship 7* – 20 February
Crew: John H. Glenn
Duration: 4 hr 55 min

Mission 6 – *Aurora 7* – 24 May
Crew: M. Scott Carpenter
Duration: 4 hr 56 min

Mission 7 – Vostok 3 – 11 August
Crew: Andrian G. Nikolayev
Duration: 3 days 22.4 hr

Mission 8 – Vostok 4 – 12 August
Crew: Pavel R. Popovich
Duration: 2 days 23 hr

Mission 9 – *Sigma 7* – 3 October
Crew: Walter M. Schirra
Duration: 9 hr 13 min

1963

Mission 10 – *Faith 7* – 15 May
Crew: L. Gordon Cooper
Duration: 1 day 10.3 hr

Mission 11 – Vostok 5 – 14 June
Crew: Valeri F. Bykovsky
Duration: 4 days 23.1 hr

Mission 12 – Vostok 6 – 16 June
Crew: Valentina V. Tereshkova
Duration: 2 days 22.8 hr

1964

Mission 13 – Voskhod 1 – 12 October
Crew: Vladimir M. Komarov, Konstantin P. Feoktistov, Boris B. Yegorov
Duration: 1 day 17 min

1965

Mission 14 – Voskhod 2 – 18 March
Crew: Pavel I. Belyayev, Alexei A. Leonov
Duration: 1 day 2 hr

Mission 15 – Gemini 3 – 23 March
Crew: Virgil I. Grissom, John W. Young
Duration: 4 hr 53 min

Mission 16 – Gemini 4 – 3 June
Crew: James A. McDivitt, Edward H. White
Duration: 4 days 1.9 hr

Mission 17 – Gemini 5 – 21 August
Crew: L. Gordon Cooper, Charles Conrad
Duration: 7 days 22.9 hr

Mission 18 – Gemini 7 – 4 December
Crew: Frank Borman, James A. Lovell
Duration: 13 days 18.6 hr

Mission 19 – Gemini 6 – 15 December
Crew: Walter M. Schirra, Thomas P. Stafford
Duration: 1 day 1.9 hr

1966

Mission 20 – Gemini 8 – 16 March
Crew: Neil A. Armstrong, David R. Scott
Duration: 10 hr 41 min

Mission 21 – Gemini 9 – 3 June
Crew: Thomas P. Stafford, Eugene A Cernan
Duration: 3 days 20 min

Mission 22 – Gemini 10 – 18 July
Crew: John W. Young, Michael Collins
Duration: 2 days 22.8 hr

Mission 23 – Gemini 11 – 12 September
Crew: Charles Conrad, Richard F. Gordon
Duration: 2 days 23.3 hr

Mission 24 – Gemini 12 – 11 November
Crew: James A. Lovell, Edwin E. Aldrin
Duration: 3 days 22.6 hr

1967

Mission 25 – Soyuz 1 – 23 April
Crew: Vladimir M. Komarov
Duration: 1 day 2.8 hr

1968

Mission 26 – Apollo 7 – 11 October
Crew: Walter M. Schirra, Donn F. Eisele, Walter Cunningham
Duration: 10 days 20.1 hr

Mission 27 – Soyuz 3 – 26 October
Crew: Georgi T. Beregovoi
Duration: 3 days 22.8 hr

Mission 28 – Apollo 8 – 21 December
Crew: Frank Borman, James A. Lovell, William A. Anders
Duration: 6 days 3 hr

1969

Mission 29 – Soyuz 4 – 14 January
Crew: Vladimir A. Shatalov
Duration: 2 days 23.3 hr

Mission 30 – Soyuz 5 – 15 January
Crew: Boris V. Volynov, Alexei S. Yeliseyev, Yevgeni V. Khrunov
Duration: 3 days 54 min

Mission 31 – Apollo 9 – 3 March
Callsigns: CSM ***Gumdrop***, LM ***Spider***
Crew: James A. McDivitt, David R. Scott, Russell L. Schweickart
Duration: 10 days 1 hr

Mission 32 – Apollo 10 – 18 May
Callsigns: CSM ***Charlie Brown***, LM ***Snoopy***
Crew: Thomas P. Stafford, John W. Young, Eugene A. Cernan
Duration: 8 days 3 min

Mission 33 – Apollo 11 – 16 July
Callsigns: CSM ***Columbia***, LM ***Eagle***
Crew: Neil A. Armstrong, Michael Collins, Edwin E. Aldrin
Duration: 8 days 3.3 hr

Mission 34 – Soyuz 6 – 11 October
Crew: Georgi S. Shonin, Valeri N. Kubasov
Duration: 4 days 22.7 hr

Mission 35 – Soyuz 7 – 12 October
Crew: Anatoli V. Filipchenko, Vladislav N. Volkov, Viktor V. Gorbatko
Duration: 4 days 22.7 hr

Mission 36 – Soyuz 8 – 13 October
Crew: Vladimir A. Shatalov, Alexei S. Yeliseyev
Duration: 4 days 22.8 hr

Mission 37 – Apollo 12 – 14 November
Callsigns: CSM ***Yankee Clipper***, LM ***Intrepid***
Crew: Charles Conrad, Richard F. Gordon, Alan L. Bean
Duration: 10 days 4.6 hr

1970

Mission 38 – Apollo 13 – 11 April
Callsigns: CSM ***Odyssey***, LM ***Aquarius***
Crew: James A. Lovell, John L. Swigert, Fred W. Haise
Duration: 5 days 22.9 hr

Mission 39 – Soyuz 9 – 1 June
Crew: Andrian G. Nikolayev, Vitali I. Sevastyanov
Duration: 17 days 17 hr

1971

Mission 40 – Apollo 14 – 31 January
Callsigns: CSM ***Kitty Hawk***, LM ***Antares***
Crew: Alan B. Shepard, Stuart A. Roosa, Edgar D. Mitchell
Duration: 9 days 2 min

Mission 41 – Soyuz 10 – 23 April
Crew: Vladimir A. Shatalov, Alexei S. Yeliseyev, Nikolai N. Rukavishnikov
Duration: 1 day 23.7 hr

Mission 42 – Soyuz 11 – 6 June
Crew: Georgi T. Dobrovolsky, Vladislav N. Volkov, Viktor I. Patsayev
Duration: 23 days 18 hr

Mission 43 – Apollo 15 – 26 July
Callsigns: CSM ***Endeavour,*** LM ***Falcon***
Crew: David R. Scott, Alfred M. Worden, James B. Irwin
Duration: 12 days 7 hr

1972

Mission 44 – Apollo 16 – 16 April
Callsigns: CSM ***Casper***, LM ***Orion***
Crew: John W. Young, Thomas K. Mattingly, Charles M. Duke
Duration: 11 days 1 hr

Mission 45 – Apollo 17 – 7 December
Callsigns: CSM ***America***, LM ***Challenger***
Crew: Eugene A. Cernan, Ronald E. Evans, Harrison H. Schmitt
Duration: 12 days 13 hr

1973

Mission 46 – Skylab 2 – 25 May
Crew: Charles Conrad, Joseph P. Kerwin, Paul J. Weitz
Duration: 28 days 49 min

Mission 47 – Skylab 3 – 28 July
Crew: Alan L. Bean, Owen K. Garriott, Jack R. Lousma
Duration: 59 days 11 hr

Mission 48 – Soyuz 12 – 27 September
Crew: Vasili G. Lazarev, Oleg G. Makarov
Duration: 1 day 23 hr

Mission 49 – Skylab 4 – 16 November
Crew: Gerald P. Carr, Edward G. Gibson, William R. Pogue
Duration: 84 days 1 hr

Mission 50 – Soyuz 13 – 18 December
Crew: Pyotr H. Klimuk, Valentin V. Lebedev
Duration: 7 days 20 hr

1974

Mission 51 – Soyuz 14 – 3 July
Crew: Pavel R. Popovich, Yuri P. Artyukhin
Duration: 15 days 17 hr

Mission 52 – Soyuz 15 – 26 August
Crew: Gennadi V. Sarafanov, Lev S. Demin
Duration: 2 days 12 min

Mission 53 – Soyuz 16 – 2 December
Crew: Anatoli V. Filipchenko, Nikolai N. Rukavishnikov
Duration: 5 days 22 hr

1975

Mission 54 – Soyuz 17 – 11 January
Crew: Alexei A. Gubarev, Georgi M. Grechko
Duration: 29 days 13 hr

Mission 55 – Soyuz 18 – 1 – 5 April
Crew: Vasili G. Lazarev, Oleg G. Makarov
Duration: 21 min 27 sec

Mission 56 – Soyuz 18 – 24 May
Crew: Pyotr I. Klimuk, Vitali I. Sevastyanov
Duration: 62 days 23 hr

Mission 57 – ASTP – 15 July
Crew: Alexei A. Leonov, Valeri N. Kubasov
Duration: 5 days 22hr

Mission 58 – ASTP – 15 July
Crew: Thomas P. Stafford, Vance D. Brand, Donald K. Slayton
Duration: 9 days 1 hr

1976

Mission 59 – Soyuz 21 – 6 July
Crew: Boris V. Volynov, Vitali M. Zholobov
Duration: 49 days 6 hr

Mission 60 – Soyuz 22 – 15 September
Crew: Valeri F. Bykovsky, Vladimir V. Aksyonov
Duration: 7 days 21 hr

Mission 61 – Soyuz 23 – 14 October
Crew: Vyacheslav D. Zudov, Valeri I. Rozhdestvensky
Duration: 2 days 6 min

1977

Mission 62 – Soyuz 24 – 7 February
Crew: Viktor V. Gorbatko, Yuri N. Glazkov
Duration: 17 days 17 hr

Mission 63 – Soyuz 25 – 9 October
Crew: Vladimir V. Kovalyonok, Valeri V. Ryumin
Duration: 2 days 44 min

Mission 64 – Soyuz 26 – 10 December
Crew: Yuri V. Romanenko, Georgi M. Grechko
Duration: 96 days 10 hr

1978

Mission 65 – Soyuz 27 – 10 January
Crew: Vladimir A. Dzhanibekov, Oleg G. Makarov
Duration: 5 days 22 hr

Mission 66 – Soyuz 28 – 2 March
Crew: Alexei A. Gubarev, Vladimir Remek
Duration: 7 days 22 hr

Mission 67 – Soyuz 29 – 15 June
Crew: Vladimir V. Kovalyonok, Alexander S. Ivanchenkov
Duration: 139 days 14 hr

Mission 68 – Soyuz 30 – 27 June
Crew: Pyotr I. Klimuk, Miroslaw Hermaszewski
Duration: 7 days 22 hr

Mission 69 – Soyuz 31 – 26 August
Crew: Valeri F. Bykovsky, Sigmund Jahn
Duration: 7 days 20 hr

1979

Mission 70 – Soyuz 32 – 25 February
Crew: Vladimir A. Lyakhov, Valeri V. Ryumin
Duration: 175 days 35 min

Mission 71 – Soyuz 33 – 10 April
Crew: Nikolai N. Rukavishnikov, Georgi I. Ivanov
Duration: 1 day 23 hr

1980

Mission 72 – Soyuz 35 – 9 April
Crew: Leonid I. Popov, Valeri V. Ryumin
Duration: 184 days 20 hr

Mission 73 – Soyuz 36 – 26 May
Crew: Valeri N. Kubasov, Bertalan Farkas
Duration: 7 days 20 hr

Mission 74 – Soyuz T-2 – 5 June
Crew: Yuri V. Malyshev, Vladimir V. Aksyonov
Duration: 3 days 22 hr

Mission 75 – Soyuz 37 – 23 July
Crew: Viktor V. Gorbatko, Pham Tuan
Duration: 7 days 20 hr

Mission 76 – Soyuz 38 – 18 September
Crew: Yuri V. Romanenko, Arnaldo T. Mendez
Duration: 7 days 20 hr

Mission 77 – Soyuz T-3 – 27 November
Crew: Leonid D. Kizim, Oleg G. Makarov, Gennadi M. Strekalov
Duration: 12 days 19 hr

1981

Mission 78 – Soyuz T-4 – 12 March
Crew: Vladimir V. Kovalyonok, Viktor P. Savinykh
Duration: 74 days 17 hr

Mission 79 – Soyuz 39 – 22 March
Crew: Vladimir A. Dzhanibekov, Jugderdemidyin Gurragcha
Duration: 7 days 20 hr

Mission 80 – STS-1 – 12 April
Orbiter Columbia
Crew: John W. Young, Robert L. Crippen
Duration: 2 days 6 hr

Mission 81 – Soyuz 40 – 15 May
Crew: Leonid I. Popov, Dumitru Prunariu
Duration: 7 days 20 hr

Mission 82 – STS-2 – 12 November
Orbiter: Columbia
Crew: Joseph H. Engle, Richard H. Truly
Duration: 2 days 6 hr

1982

Mission 83 – STS-3 – 22 March
Orbiter: Columbia
Crew: Jack R. Lousma, Charles G. Fullerton
Duration: 8 days 4 min

Mission 84 – Soyuz T-5 – 13 May
Crew: Anatoli Berezovoi, Valentin V. Lebedev
Duration: 211 days 9 hr

Mission 85 – Soyuz T-6 – 24 June 1982
Crew: Vladimir A. Dzhanibekov, Alexander S. Ivanchenkov, Jean-Loup Chretien
Duration: 7 days 21 hr

Mission 86 – STS-4 – 27 June
Orbiter: Columbia
Crew: Thomas K. Mattingly, Henry W. Hartsfield
Duration: 7 days 1 hr

Mission 87 – Soyuz T-7 – 19 August
Crew: Leonid I. Popov, Alexander A. Serebrov, Svetlana Y. Savitskaya
Duration: 7 days 21 hr

Mission 88 – STS-5 – 11 November
Orbiter: Columbia
Crew: Vance D. Brand, Robert F. Overmyer, William B. Lenoir, Joseph P. Allen
Duration: 5 days 2 hr

1983

Mission 89 – STS-6 – 4 April
Orbiter: Challenger
Crew: Paul J. Weitz, Karol J. Bobko, F. Story Musgrave, Donald H. Peterson
Duration: 5 days 24 min

Mission 90 – Soyuz T-8 – 20 April
Crew: Vladimir G. Titov, Gennadi M. Strekalov, Alexander A. Serebrov
Duration: 2 days 17 min

Mission 91 – STS-7 – 18 June
Orbiter: Challenger
Crew: Robert L. Crippen, Frederick H. Hauck, John M. Fabian, Sally K. Ride, Norman E. Thagard
Duration: 6 days 2 hr

Mission 92 – Soyuz T-9 – 27 June
Crew: Vladimir A. Lyakhov, Alexander P. Alexandrov
Duration: 149 days 10 hr

Mission 93 – STS-8 – 30 August
Orbiter: Challenger
Crew: Richard H. Truly, Daniel C. Brandenstein, Dale A. Gardner, Guion S. Bluford, William E. Thornton
Duration: 6 days 1 hr

Mission 94 – STS-9 – 28 November
Orbiter: Columbia
Crew: John W. Young, Brewster H. Shaw, Owen K. Garriott, Robert A. R. Parker, Ulf Merbold, Byron K. Lichtenberg
Duration: 10 days 7 hr

1984

Mission 95 – STS 41-B – 3 February
Orbiter: Challenger
Crew: Vance D. Brand, Robert L. Gibson, Bruce McCandless, Ronald E. McNair, Robert L. Stewart
Duration: 7 days 23 hr

Mission 96 – Soyuz T-10 – 8 February
Crew: Leonid D. Kizim, Vladimir A. Solovyov, Oleg Y. Atkov
Duration: 236 days 22 hr

Mission 97 – Soyuz T-11 – 3 April
Crew: Yuri V. Malyshev, Gennadi M. Strekalov, Rakesh Sharma
Duration: 7 days 21 hr

Mission 98 – STS 41-C – 6 April
Orbiter: Challenger
Crew: Robert L. Crippen, Francis R. Scobee, Terry J. Hart, George D. Nelson, James D. A. van Hoften
Duration: 6 days 23 hr

Mission 99 – Soyuz T-12 – 17 July
Crew: Vladimir A. Dzhanibekov, Svetlana Y. Savitskaya, Igor P. Volk
Duration: 11 days 19 hr

Mission 100 – STS 41-D – 30 August
Orbiter: Discovery
Crew: Henry W. Hartsfield, Michael L. Coats, Steven A. Hawley, Judith A. Resnik, R. Michael Mullane, Charles D. Walker
Duration: 6 days 56 min

Mission 101 – STS 41-G – 5 October
Orbiter: Challenger
Crew: Robert L. Crippen, Jon A. McBride, Sally K. Ride, Kathryn D. Sullivan, David C. Leestma, Paul Scully-Power, Marc Garneau
Duration: 8 days 5 hr

Mission 102 – STS 51-A – 8 November
Orbiter: Discovery
Crew: Frederick H. Hauck, David M. Walker, Dale A. Gardner, Joseph P. Allen, Anna L. Fisher
Duration: 7 days 23 hr

1985

Mission 103 – STS 51-C – 24 January
Orbiter: Discovery
Crew: Thomas K. Mattingly, Loren J. Shriver, Ellison S. Onizuka, James F. Buchli, Gary Payton
Duration: 3 days 33 min

Mission 104 – STS 51-D – 12 April
Orbiter: Discovery
Crew: Karol J. Bobko, Donald E. Williams, S. David Griggs, Jeffrey A. Hoffman, M. Rhea Seddon, Edwin Garn, Charles D. Walker
Duration: 6 days 23 hr

Mission 105 – STS 51-B – 29 April
Orbiter: Challenger
Crew: Robert F. Overmyer, Frederick D. Gregory, Norman E. Thagard, William E. Thornton, Don L. Lind, Lodewijk van den Berg, Taylor G. Wang
Duration: 7 days 8 min

Mission 106 – Soyuz T-13 – 6 June
Crew: Vladimir A. Dzhanibekov, Viktor P. Savinykh
Duration: Dzhanibekov 112 days; Savinykh 168 days

Mission 107 – STS 51-G – 17 June
Orbiter: Discovery
Crew: John O. Creighton, John M. Fabian, Shannon W. Lucid, Steven R. Nagel, Patrick Baudry, Prince Sultan A. A. Al-Saud
Duration: 7 days 1 hr

Mission 108 – STS 51-F – 29 July
Orbiter: Challenger
Crew: Charles G. Fullerton, Roy D. Bridges, Anthony W. England, Karl G. Henize, F. Story Musgrave, Loren W. Acton, John-David Bartoe
Duration: 7 days 22 hr

Mission 109 – STS 51-I – 27 August
Orbiter: Discovery
Crew: Joseph H. Engle, Richard O. Covey, William F. Fisher, John M. Lounge, James D. A. van Hoften
Duration: 7 days 2 hr

Mission 110 – Soyuz T-14 – 17 September
Crew: Vladimir V. Vasyutin, Georgi M. Grechko, Alexander A. Volkov
Duration: 64 days 21 hr

Mission 104 – STS 51-J – 3 October
Orbiter: Atlantis
Crew: Karol J. Bobko, Ronald J. Grabe, David C. Hilmers, Robert L. Stewart, William Pailes
Duration: 4 days 1 hr

Mission 112 – STS 61-A – 30 October
Orbiter: Challenger
Crew: Henry W. Hartsfield, Steven R. Nagel, Guion S. Bluford, James F. Buchli, Bonnie J. Dunbar, Reinhard Furrer, Ernst W. Messerschmid, Wubbo Ockels
Duration: 7 days 44 min

Mission 113 – STS 61-B – 27 November
Orbiter: Atlantis
Crew: Brewster H. Shaw, Bryan D. O'Connor, Mary L. Cleave, Sherwood C. Spring, Jerry L. Ross, Rodolfo N. Vela, Charles D. Walker
Duration: 6 days 21 hr

1986

Mission 114 – STS 61-C – 12 January
Orbiter: Columbia
Crew: Robert L. Gibson, Charles F. Bolden, Steven A. Hawley, George D. Nelson, Franklin R. Chang-Diaz, Robert J. Cenker, C. William Nelson
Duration: 6 days 2 hr

Mission 115 – STS 51-L – 28 January
Orbiter: Challenger
Crew: Francis R. Scobee, Michael J. Smith, Judith A. Resnik, Ronald E. McNair, Ellison S. Onizuka, Gregory B. Jarvis, S. Christa McAuliffe
Duration: 73 sec

Mission 116 – Soyuz T-15 – 13 March
Crew: Leonid D. Kizim, Vladimir A. Solovyov
Duration: 125 days

1987

Mission 117 – Soyuz TM-2 – 5 February
Crew: Yuri V. Romanenko, Alexander I. Leveikin
Duration: Romanenko, 326 days; Leveikin, 177 days

Mission 118 – Soyuz TM-3 – 22 July
Crew: Alexander S. Viktorenko, Alexander P. Alexandrov, Muhammad Faris
Duration: Viktorenko and Faris 8 days; Alexandrov 160 days

Mission 119 – Soyuz TM-4 – 21 December
Crew: Vladimir G. Titov, Musakhi K. Manarov, Anatoli S. Levchenko
Duration: Levchenko 8 days; Titov and Manarov
365 days 22 hr 39 min

1988

Mission 120 – Soyuz TM-5 – 7 June
Crew: Anatoli Y. Solovyov, Viktor P. Savinykh, Alexander P. Alexandrov
Duration: 9 days 20 hr

Mission 121 – Soyuz TM-6 – 29 August
Crew: Vladimir A. Lyakhov, Valeri Poliakov, Abdul A. Mohmand
Duration: Lyakhov and Mohmand 8 days 20 hr

Mission 122 – STS-26 – 29 September
Orbiter: Discovery
Crew: Frederick H. Hauck, Richard O. Covey, John M. Lounge, David C. Hilmers, George D. Nelson
Duration: 4 days 2 hr

Mission 123 – Soyuz TM-7 – 26 November
Crew: Alexander A. Volkov, Sergei Krikalev, Jean-Loup Chretien
Duration: Chretien 24 days 18 hr

Mission 124 – STS-27 – 2 December
Orbiter: Atlantis
Crew: Robert L. Gibson, Guy S. Gardner, Jerry L. Ross, R. Michael Mullane, William M. Shepherd
Duration: 4 days 9 hr

GLOSSARY

ablation The melting and boiling away of a heat shield during re-entry, a process that dissipates the heat produced by air friction.
abort Cut short a flight or a mission.
acquisition Making contact with a spacecraft so that information can be transmitted to, or received from, it.
aerodynamic vehicle A vehicle able to fly in the atmosphere, as opposed to a space vehicle, which normally cannot.
aerospace A word coined from 'aeronautics' and 'space'. Aerospace craft are those like the shuttle, designed to operate in the air as well as in space.
airlock A chamber in a spacecraft which can be pressurized with air and depressurized, through which astronauts leave and enter their craft when they go spacewalking.
ALSEP The Apollo Lunar Surface Experiments Package, which formed the basis of the automatic scientific stations the Apollo astronauts set up on the Moon.
AM The airlock module on Skylab.
antimatter A form of matter in which the atoms are composed of antiparticles – particles like protons and electrons, but with opposite electric charge, ie positive electrons and negative protons.
apogee The most distant point a satellite reaches in its orbit around the Earth.
apogee motor A rocket fitted to a spacecraft which is fired at the high point of an elliptical parking orbit to place it in geostationary orbit.
APU Auxiliary power unit.
artificial gravity A force similar to the force of gravity, created in a space station, say, by rotating the station on its axis.
artificial satellite A man-made object put into orbit around the Earth, or other planet or moon. Usually just termed satellite, though strictly 'satellite' refers to a natural body, such as the Earth's Moon.
ASAT An anti-satellite satellite.
ASTP The Apollo-Soyuz Test Project, the joint flight in July 1975 of Apollo and Soyuz astronauts and cosmonauts.
asteroids Or minor planets; small rocky bodies in the solar system that circle the Sun mainly in a broad band (the asteroid belt) between the orbits of Mars and Jupiter.
astrobiology See **exobiology.**
astronautics The science of space travel; hence **astronaut**, a space traveler.
astronomical unit The distance between the Earth and the Sun, approximately 150 million km (93 million miles)
astronomy The scientific study of the heavens.
ATM The Apollo telescope mount, one of the modules of Skylab.
atmosphere The layer of gases surrounding a planet or a moon; the Earth's atmosphere is composed mainly of nitrogen and oxygen. But the atmospheres of other planets are composed of a totally different gas mixture – that of Venus, for example, is mainly carbon dioxide.
attitude The position of a craft in relation to something else, such as the horizon.
aurora A display of colored lights and streamers seen mainly in polar regions, caused by particles in the solar wind interacting with gases in the upper atmosphere. In the northern hemisphere the phenomenon is called the aurora borealis; in the southern, the aurora australis.
avionics The electronic systems and instruments that monitor and control a space vehicle or plane.
back-up A person or item of equipment available to take over the function of another person or item should the need arise. For many space missions there are prime and back-up crews.
ballistic missile A missile guided only in its launch phase, then arcing up and over in a ballistic trajectory towards its target.
barbecue mode Slowly rolling a spacecraft or space station to prevent it becoming overheated by the Sun.
biosensors Sensors stuck on the body to measure biological functions, such as heart beat and respiration rate.
blackout The loss in communications with a spacecraft that is re-entering the atmosphere, caused by ionization of the surrounding air. Medically a blackout is loss of consciousness that can occur when a person undergoes high g-forces.
boilerplate A metal replica of a spacecraft flown during early tests of the vehicle.
booster The first stage of a launch vehicle or additional rocket stages attached to the core vehicle to give extra thrust at lift-off.
burn The period during which a rocket fires.
burn-up The burning up and disintegration of a satellite when re-entering the atmosphere.
CapCom Abbreviation for Capsule Communicator, the person at Mission Control who communicates with astronauts in space.
capsule The name given to the cramped crew cabin of early spacecraft such as Mercury and Vostok.
celestial Relating to the heavens.
celestial mechanics Study of the mechanisms governing the motions in space of the heavenly bodies, satellites and probes.
centrifuge A training aid for astronauts, a machine that whirls them round in a capsule at the end of a long arm, subjecting them to centrifugal forces that duplicate the g-forces they will experience on lift-off and re-entry.
Clarke orbit Same as geostationary orbit, 35,900 km (22,300 miles) high; named after space writer Arthur C. Clarke, who first suggested its use for communications satellites in 1948.
combustion chamber The part of a rocket in which the propellants are burned.
CM Command module of the Apollo spacecraft.
CNES Centre National d'Etudes Spatiales, France's center for space research.
communications satellite One that relays telecommunications signals – telephone, telex, fax, computer data – between ground stations.
comsat A communications satellite.
cosmic rays Charged particles that come from outer space, mainly protons and electrons.
cosmodrome A launch center for Soviet spacecraft; a spaceport.
cosmonaut The Soviet term for a space traveler.
cosmos An alternative term for the universe, space.
COSPAS A series of Soviet satellites equipped to monitor emergency radio frequencies used by aircraft and shipping in distress. They operate in conjunction with US sarsats.
countdown The counting down to zero of a certain period of time before a space launch.
cryogenic propellants Propellants like liquid hydrogen and liquid oxygen, both of which exist at very low temperatures (−253°C and −183°C respectively).
CSM The combined command and service modules of the Apollo spacecraft.
decay The gradual lowering of a satellite's orbit caused by atmospheric drag slowing it down.
de-orbit Come out of orbit; usually effected by a de-orbit burn, a fire of retrorockets.
deploy Place in position.
docking The joining together of two spacecraft in space.
DOD Department of Defense.
drag The resistance experienced by a body traveling through the air.
drogue A small parachute that deploys first to stabilize a spacecraft when it comes in to land, preceding deployment of the main parachutes.
DSN NASA's Deep Space Network, a communications and tracking network for distant space probes operated by Jet Propulsion Laboratory in California.
ELV Expendable launch vehicle – one that can be used only once.
EMU Extravehicular mobility unit, the correct term for the space shuttle spacesuit.
encounter A meeting between a space probe and its target.
equatorial orbit An orbit in the plane of the equator.
EROS Earth Resources Observation System. The EROS Data Center is the prime center in the US for processing Landsat data and distributing Landsat imagery.
ESA The European Space Agency, the body responsible for coordinating space activities in Europe.

escape velocity The speed a body must possess to escape completely from Earth's gravity; about 40,000 km/h (25,000 mph).

ET External tank, of the shuttle; also short for extraterrestrial.

Eumetsat The European Meteorological Satellite Organization.

Eutelsat The European Telecommunications Satellite Organization.

EVA Extravehicular activity. Activity outside a spacecraft; commonly called spacewalking.

exobiology The study relating to the possibility of life outside the Earth, on other planets or in other solar systems.

expendable Can be used only once, as in expendable launch vehicle.

extraterrestrial Existing outside the Earth.

eyeballs in A term that relates to the effect of the g-forces generated by fierce acceleration on the human body.

footprint The area on the ground over which signals can be received from a communications satellite; or the area over which debris from a disintegrating satellite can be scattered after re-entry.

fly-by A space mission in which a probe flies past a planet or moon without going into orbit around it or landing.

free fall The state that exists in orbit when everything is 'falling around the Earth'. It gives rise to the phenomenon we know better as weightlessness.

fuel cell An electric cell that produces electricity by combining hydrogen and oxygen gases to form water; used to supply electrical power in the Apollo spacecraft and the space shuttle.

g The symbol for the acceleration due to gravity. We can say the force on our bodies due to gravity is 1g; see **g-forces.**

gantry A tower that gives access to a launch vehicle on the launch pad.

geostationary orbit An orbit 35,900 km (22,300 miles), in which a satellite circles the Earth once every 24 hours and therefore appears to be fixed in the sky.

getaway special An experiment carried in a small container in the space shuttle. The getaway special program gives universities and other organizations access to space at reasonable cost.

g-forces The forces a body experiences when it is subjected to high acceleration or deceleration, eg during a rocket lift-off or spacecraft re-entry. 3g, for example, means that the forces experienced are three times the normal pull of gravity.

gimbal A pivoting device which allows a rocket nozzle, for example, to swivel, permitting directional control during firing.

glitch A problem of some sort.

global positioning system A US satellite navigation system for shipping, consisting of 18 Navstar satellites.

gravity The pull the Earth exerts on everything on and around it in space. Every massive body exerts a similar gravitational attraction, the strength of which depends on its mass. Small bodies like the Moon have less gravitational pull.

gravity-assist A method that uses the gravitational attraction of a planet to increase the speed of a space probe.

ground station Also Earth station; a base, usually equipped with directional dish antennae, for relaying radio signals to satellites and probes.

g-suit Clothing worn by astronauts and pilots that applies pressure on the lower body and legs to prevent blood collecting there under high g-forces.

hatch A door in the hull of a spacecraft with an airtight seal.

heat shield A coating on the outside of a spacecraft designed to protect it and any astronauts inside from the heat developed during re-entry. Both the US and Soviet shuttles use ceramic tiles as a heat shield.

hold A temporary halt in a countdown.

housekeeping Routine activities performed on-board a spacecraft by computers and crew.

hypergolic propellants Propellants that set alight when they mix.

hypersonic Traveling at more than five times the speed of sound.

ICBM Intercontinental ballistic missile; a long-range ballistic missile able to travel between continents. Modified ICBMs were used to launch the first satellites.

ignition The instant when a rocket engine is started, when the propellants are ignited. Thereafter the propellants continue to burn by themselves.

inertial guidance A built-in guidance system fitted to launch vehicles, which uses gyroscopes and accelerometers on three axes to keep the craft to a preprogramed trajectory.

inject Boost a spacecraft into a particular orbit.

Inmarsat The International Maritime Satellite Organization.

inner planets The planets Mercury, Venus, Earth and Mars that lie comparatively near to the Sun. Also called the terrestrial planets. Compare **outer planets.**

Intelsat The International Telecommunications Satellite Organization.

interplanetary Between the planets.

interstellar Between the stars.

IUS Inertial upper stage; a two-stage booster rocket used on shuttle flights to boost payloads into high Earth-orbits or to the planets.

jettison Discard. Parts of a spacecraft may be jettisoned before re-entry.

Kosmolyot The name once given to the Soviet space shuttle; now called *Buran (Snowstorm)*.

Lagrangian points Regions in space where the gravitational influences of two massive bodies, such as the Earth and Moon, can hold smaller bodies in a stable orbit. Future space colonies may be built at such points.

launch vehicle A system of rockets that can propel a payload into space; a step rocket.

launch window The period of time during which a spacecraft can be launched so it can reach its planned target or objective.

life-support system The system in a spacecraft that keeps the astronaut crew alive and in a comfortable environment.

light-year The distance light travels in a year, nearly 10 million million km (6 million million miles), which provides a useful unit of measurement in astronomy.

LM Sometimes LEM; the Apollo lunar module.

LOR Lunar orbit rendezvous; the method by which Apollo reached and landed on the Moon.

LOX Liquid oxygen.

LRV Lunar roving vehicle; the Moon buggy the Apollo astronauts used when exploring the Moon.

lunar Relating to the Moon.

lunarnaut An astronaut who travels to or walks on the Moon.

Mach number The speed of a craft through the air compared with the local speed of sound.

magnetic field The region around a body where its magnetism acts. The Sun and some of the planets, including the Earth, have a magnetic field.

magnetosphere The great 'bubble' that extends way out into space around a heavenly body with a magnetic field where the solar wind interacts with the magnetic field. The magnetosphere is irregular in shape, rather like a raindrop, with the blunt end facing into the solar wind.

manned maneuvering unit See **MMU.**

man-rating Developing and testing a rocket or space vehicle so that it is safe to be flown by astronauts.

mare A lunar sea; plural maria.

mascon Areas of the Moon of unusually high gravity where there appear to be concentrations of high-density matter; found particularly in the maria.

microgravity The minute amount of gravity existing between objects in orbit because of their mass. Compared with the gravity of the Earth, which is effectively cancelled in orbit, microgravity is a very very weak force, but it can nevertheless be exploited in experiments.

MCC Mission Control Center.

MDA Multiple docking adapter; one of the Skylab modules.

MET Mission elapsed time, ie time from lift-off.

meteor The streak of light we see in the night sky caused by particles of rock from outer space burning up due to air friction in the upper atmosphere.
meteorite Rock from outer space that has survived entry through the atmosphere and fallen to the ground.
meteoroid A particle of dust or rocky matter traveling in space, which we see as a meteor when it hits the Earth's atmosphere.
mid-course correction A rocket burn made during a lengthy flight to adjust the speed of a spacecraft so that it reaches its target.
mission A space flight.
mission control The communications and control center for a space mission. The US mission control center for manned space flight is located at the Johnson Space Center, Houston, Texas. US satellite mission control is located at the Goddard Space Flight Center, Greenbelt, Maryland; mission control for planetary explorations is at the Jet Propulsion Laboratory, Pasadena, California. The main Soviet mission control center is located at Kaliningrad on the Baltic Sea coast due west of Moscow. For the European Space Agency, mission control is at Darmstadt, West Germany.
mission specialist A shuttle astronaut with specific duties related to a particular flight, especially with regard to the payload.
MMU Manned maneuvering unit; a jet-propelled backpack used by shuttle astronauts for moving about in space.
mock-up A full-size dummy spacecraft that looks like the real thing, often fitted with instruments and controls, which may be 'live'. See also **simulator.**
module A major self-contained unit of a spacecraft.
monitor Keep an eye on; keep track of.
moonquake Ground tremors in the Moon's crust, caused by internal activity or the impact of meteorites.
MSS Multispectral scanner.
multispectral scanner MSS; a sensor on Landsat satellites that scans the ground at various wavelengths.
multistage rocket Another name for a step rocket.
NASA National Aeronautics and Space Administration; the body that organizes aerospace activities in the United States.
NASDA Japan's National Space Development Agency.
navsat A navigation satellite.
newton The unit in which the thrust of a rocket is often expressed. It is the force that will give a mass of 1 kg an acceleration of 1 metre per second. Named after Isaac Newton.
NOAA National Oceanic and Atmospheric Administration; the organization that manages the US weather and Earth-resources satellite programs.
nominal A term which means that everything is as it should be.
NORAD The North American Air-Defense Command, which constantly monitors all satellites (and debris) in orbit.
nose cone The streamlined front section of a rocket that reduces drag as the rocket passes through the atmosphere.
nozzle The bell-shaped section at the rear of a rocket, which allows the hot gases coming from the combustion chamber to expand with the optimum efficiency.
OMS Orbital maneuvering system, of the shuttle; two engines that fire to inject the vehicle into orbit and to bring about de-orbit.
OMV Orbital maneuvering vehicle; a robot rocket vehicle used to move satellites and other units in orbit; being developed for use with the space station *Freedom.*
orbit The path in space taken by one body traveling around another. Orbits are usually elliptical.
orbital maneuvering system See **OMS.**
orbital manoeuvering vehicle See **OMV.**
orbital period The time it takes a satellite to complete one orbit. For a satellite orbiting a few hundred kilometers above the Earth, the orbital period is about 1½ hours.
orbital transfer vehicle See **OTV.**
orbital velocity The speed a satellite needs to remain in orbit. Around the Earth at an altitude of about 300 km (200 miles), the orbital velocity is about 28,000 km/h (17,500 mph).
orbiter A spacecraft designed to orbit a planet or moon; specifically the space shuttle orbiter.
OTV Orbital transfer vehicle; a reusable rocket-powered robot vehicle designed to boost satellites from low to high orbits; to be operated in conjunction with space station *Freedom*.
outer planets The planets Jupiter, Saturn, Uranus, Neptune and Pluto, which lie in the outer solar system. Compare **inner planets.**
OWS Orbital workshop; the main part of Skylab, built out of a redundant SIVB rocket stage.
oxidizer The propellant in a rocket that provides the oxygen to burn the fuel.
parking orbit An orbit around the Earth in which a spacecraft 'parks' temporarily before being boosted into its final orbit or trajectory.
PAM Payload assist module; a booster rocket stage attached to a payload to boost it into a high orbit; used on the space shuttle and the Delta rocket.
payload The cargo a launch vehicle or rocket carries.
payload specialist A shuttle astronaut who goes into space, probably only once, to supervise the operation of a certain experimental payload.
perigee The point in a satellite's orbit when it is closest to the Earth.
photon A 'packet', or quantum of light energy.
pitching The up-and-down oscillation of a spacecraft's nose (and tail).
PLSS Portable life-support system; the self-contained backpack astronauts wear when they go spacewalking.
pogo The up and down vibrations in a rocket structure at launch generated by the thrust of the engines.
polar orbit A satellite orbit that takes it over the North and South Poles.
pressure suit A suit and helmet which is pressurized with air to enable pilots and astronauts to survive at high altitudes or in space.
probe A spacecraft that escapes from the Earth's gravity and travels to the Moon or the planets.
propellant A substance burned in a rocket engine to develop thrust to propel it. Two propellants are generally needed, a fuel (such as liquid hydrogen or kerosene) and ar oxidizer (such as liquid oxygen).
radio telescope A telescope designed to gather the faint radio waves that heavenly bodies give out; sometimes used to receive data from distant probes.
RCS Reaction control system; a system of thrusters on a spacecraft fired to change its attitude in space.
redundancy The duplication of vital parts in a system, so that if one fails, another can take over.
reaction principle Stated by Isaac Newton, it is: to every action, there is an equal and opposite reaction. In other words, when a force acts in one direction, there is always an equal force acting in the opposite direction.
re-entry The moment when a spacecraft or vehicle hits the atmosphere when returning from space. The drag, or resistance of the air, slows the craft down but causes it to heat up due to air friction.
regenerative cooling The method of cooling a rocket engine by circulating cold incoming fuel through a double wall in the engine nozzle.
regolith The finely divided 'soil' that covers the Moon's surface.
remote sensing The gathering of data and imagery from a distance; especially of the Earth's surface from satellites, such as Landsat and SPOT.
rendezvous The meeting of two craft, eg in orbit prior to docking.
rescue ball A flexible ball container, equipped with its own life-support system, in which shuttle astronauts lacking a spacesuit could be transferred to another craft during a rescue mission.
retrofire Firing a rocket backwards, ie in the direction of travel, to act as a brake; hence retrobraking.
RMS Remote manipulator

system, of the shuttle; the shuttle's 'crane', a remote-controlled jointed arm fitted into the orbiter's payload bay.
rocket An engine that produces a stream of hot gases for propulsion, working on the reaction principle. It can work in airless space because it carries not only fuel, but also the oxygen to burn the fuel.
rolling An oscillating rotational movement of a spacecraft about a lengthwise axis.
rollout The time when a space vehicle is moved to the launch pad prior to launch.
rookie A novice.
RTG Short for radioisotope thermoelectric generator, a nuclear power source that converts the heat produced by the decay of a radioactive isotope into electricity to power a spacccraft (cg Voyager).
satellite A small body that orbits around a larger one in space, such as the Moon around the Earth. The Moon is Earth's natural satellite. The term satellite these days invariably means artificial satellite, a man-made moon.
sarsat A search and rescue satellite, equipped to relay signals from emergency beacons on-board planes and ships; see also **COSPAS.**
SDI Strategic Defense Initiative; see **Star Wars.**
SETI Search for extraterrestrial intelligence; a program that uses radio telescopes to listen out for signals from space that could announce the presence out there of other intelligent civilizations.
shirt-sleeve environment One in which people can wear ordinary clothing, with no need for pressure suits or spacesuits.
shock diamonds A series of diamond-shaped patterns seen in a rocket exhaust created by shock waves; they can be seen in the shuttle main engine exhaust.
simulator A full-scale mock-up of a spacecraft which simulates the behavior of the real craft.
SM Service module; part of the Apollo spacecraft.
solar Relating to the Sun.
solar cell A silicon-wafer device that produces electricity when sunlight falls on it.
solar sail A method proposed to power long-distance spacecraft by means of large 'sails', which would harness the momentum of the photons which make up sunlight.
solar system The Sun and the family of planets, moons, asteroids, meteors and comets that circle around it.
solar wind The stream of charged particles emitted into space by the Sun.
space colony A habitat built in space that could provide a permanent home for thousands of people.
space medicine The study of the human body in space, particularly with regard to the effects on it of the weightless environment.
space sickness The nausea experienced by many astronauts during the early part of a space mission. The condition is properly called space adaptation syndrome.
spacesuit A multilayer garment astronauts wear in space as protection against the vacuum, cold and harmful radiations.
space tug Another name for the OMV.
spacewalk The popular name for EVA (extravehicular activity).
speed of light In a vacuum light travels at a speed of about 300,000 km (186,000 miles) per second. Radio waves and all other electromagnetic waves also travel at this speed, which, according to relativity theory, is the highest speed at which anything can travel.
spin stabilization Spinning a spacecraft about an axis to keep it in a stable orientation; this takes advantage of the principle of the gyroscope.
splashdown The landing of a spacecraft at sea.
SPOT Satellite Probatoire pour l'Observation de la Terre; a French Earth resources satellite.
SRB Solid rocket booster; two are used on the shuttle.
stage One of the rocket sections of a step rocket.
starship A spaceship designed to travel through interstellar space to visit the planets of other solar systems.
Star Wars The popular name given to the proposed space-based defense system, the Strategic Defense Initiative (SDI), designed to destroy oncoming missiles while they are still in space.
stellar Relating to the stars.
step rocket A rocket made up of several rocket units (stages) joined together, usually end to end.
STS Space Transportation System; the correct name for the US space shuttle system. Shuttle flights are designated a STS number.
suborbital A space flight in which a craft does not go into orbit, but follows a ballistic trajectory.
supersonic Traveling at speeds greater than the speed of sound.
TDRS Tracking and data relay satellite; a powerful communications satellite in geostationary orbit used to relay tracking information, communications and data from low-flying satellites, particularly the space shuttle, into NASA's communications network.
telemetry Transmitting instrument readings between a spacecraft and ground control.
terminator The boundary between light and shadow – day and night – on a planet or moon.
terrestrial Relating to the Earth.
thematic mapper A scanning instrument on the Landsat satellite that scans at seven wavelengths.
thrust The pushing force that propels a rocket, measured in newtons (SI units) or pounds.
thruster A small rocket engine used for maneuvering.
touchdown The moment of landing of a a returning shuttle or a space probe.
tracking Following the path of a spacecraft through space by radio, radar or photography.
tracking and data relay satellite See **TDRS.**
trajectory The flight path of a body.
transfer orbit The elliptical path taken by a spacecraft changing from one orbit to another.
UFO Unidentified flying object; one whose appearance and behavior in the sky cannot readily be explained. Popularly, UFOs are thought to be spaceships from another world.
umbilical A connecting line carrying power, air, water, oxygen between, say, a launch-pad gantry and a launch vehicle; or an astronaut and his spacecraft.
V1, V2 Hitler's vengeance (Vergeltungswaffe) weapons of the Second World War. The V1 was a winged flying bomb, known as the Buzz bomb because of its rasping pulse-jet engine. The V2 was a rocket missile that pioneered a terrible new type of warfare and was the direct ancestor of today's space launch vehicles.
VAB Vehicle Assembly Building; the massive building at Kennedy launch complex 39, built for the assembly of Saturn V Moon rockets and now used for assembling the space shuttle.
vernier A rocket engine that gives tiny thrusts.
VKK VozdushnoKosmichesiy-Korabl; formal name of the Soviet space shuttle system; equivalent to the American STS.
weightlessness The condition astronauts experience in orbit, when their bodies appear not to have any weight. The proper term for this state is free fall.
yawing The side-to-side oscillation of a spacecraft's nose (and tail).
zero-g Another term for weightlessness.

INDEX

(Numbers in *italics* refer to illustrations)

A

B

C

D

E

F

G

H

I

J

K

L

M

N

O

PQ

R

S

T

UV

W

XYZ

Acknowledgements/ Credits

The author would like to extend his grateful thanks for their unstinting advice and encouragement to his many friends at NASA Headquarters in Washington DC; the Kennedy Space Center at Cape Canaveral, Florida; the Johnson Space Center at Houston, Texas; the Goddard Space Flight Center at Greenbelt, Maryland; the Jet Propulsion Laboratory at Pasedena, California; the Alabama Space and Rocket Center at Huntsville, Alabama; to the staff at the Headquarters of the European Space Agency (ESA) in Paris; the Novosti Press Agency in London; and the National Space Development Agency (NASDA) in Tokyo.

The majority of the photographs illustrating this book have been provided by NASA, the European Space Agency and the Novosti Press Agency through Spacecharts Photo Library, to whom many thanks. Thanks are also due to the following for providing pictures: British Interplanetary Society 47(BR); EROS 31; Imperial War Museum 14(L); NASDA 20(T); US Naval Observatory 18(L); Royal Aircraft Establishment Farnborough 29, 33. The photographs on the following pages were taken by the author: 39, 46, 47(BL), 52(B), 53(R), 54(TL), 70, 79(B), 99(T), 108.